VIRTUAL LANs

THE McGRAW-HILL SERIES ON COMPUTER COMMUNICATIONS (SELECTED TITLES)

Virtual LANs
A Guide to Construction, Operation, and Utilization

Marina Smith

McGraw-Hill
New York   San Francisco   Washington, D.C.   Auckland   Bogotá
Caracas   Lisbon   London   Madrid   Mexico City   Milan
Montreal   New Delhi   San Juan   Singapore
Sydney   Tokyo   Toronto

Library of Congress Cataloging-in-Publication Data

Smith, Marina.
    Virtual LANs : a guide to construction, operation, and utilization /
Marina Smith.
        p.   cm. — (The McGraw-Hill series on computer communications)
    Includes index.
    ISBN 0-07-913623-0
    1. Local area networks (Computer networks)—Computer programs.
    2. Virtual computer systems.   I. Title.   II. Series.
    TK5105.7.S578   1997
    004.6'8—dc21   97-37946
                                                                              CIP

McGraw-Hill

A Division of The McGraw·Hill Companies

    3 4 5 6 7 8 9 0   DOC/DOC   9 0 2 1 0 9

ISBN 0-07-913623-0

*The sponsoring editor for this book was Steven Elliot, the editing supervisor was Bernard
Onken, and the production supervisor was Clare Stanley. It was set in Vendome by
Priscilla Beer of McGraw-Hill's Professional Book Group composition unit.*

Printed and bound by R. R. Donnelley & Sons Company.

McGraw-Hill books are available at special quantity discounts to use as premi-
ums and sales promotions, or for use in corporate training programs. For more
information, please write to the Director of Special Sales, McGraw-Hill, 11 West
19th Street, New York, NY 10011. Or contact your local bookstore.

*This book is printed on recycled, acid-free paper containing a minimum of
50% recycled de-inked fiber.*

To my parents, for always encouraging me to be a writer.

CONTENTS

vii

Contents

PREFACE

As companies worldwide change their information technology (IT) infra-structure to accommodate more flexible working practices, they are implementing new types of networking. The subject of one of these new ways, virtual local area networks (VLANs), is much bandied about at the moment. But what are they? How do they work? Do you need them in your networks? The topic is frequently discussed by marketeers, engineers, and network managers. Products are available (just about), although it is doubtful whether they work properly yet. The purpose of this book is to provide network managers and administrators and all readers who have an interest in the subject with general information as to the purpose and implementation of VLANs and detailed information on VLAN construc-tion, operation, and utilization. It aids decision making by stating the case for and against VLANs as a whole, by discussing the different types avail-able (or soon to become available), and by examining the problems that arise with them.

VLANs depends on switching, and a section on that subject will bring the reader up to speed on this latest networking technology. A back-grounder [including the OSI (Open Systems Interconnection) reference model], along with brief reminders as to the types of devices used in local area networks, will fill in any other gaps in the reader's knowledge. Then there is a detailed presentation of how VLANs are constructed and the types available. At present, given the proprietary nature of products now available, and the multiplicity of standards that have a bearing on VLANs, there is a summary of all the relevant standards.

A strong section on management of VLANs guides you through the pitfalls of implementation and is followed by an outline of the security implications of VLAN installation.

A useful feature of this work is the specific implementation notes given for all the major offerings in the field, together with a roundup of several others. The vendors are also briefly profiled.

MARINA SMITH

ACKNOWLEDGMENTS

I could not have done this book alone, even with the availability of information from the World Wide Web. Such information needs interpretation, and for that, and for a lot of help from friends, associates, and vendors, as well as from all the people who put information onto Web pages, I am truly grateful.

I have had a lot of help in the preparation of this book from consultants, vendors, Internet newsgroup correspondents, and friends. I want to thank them all, especially Ken Mann, independent consultant, for his help with the network management and security sections; John M. Wobus of Syracuse University for his DHCP (Dynamic Host Configuration Protocol) information; and James Eibisch for his help with the figures. I am also greatly indebted to Datapro Information Services for support during the preparation of this work and for permission to use parts of several reports within this book, especially to Rob McCombe.

I am also grateful to people at many of the vendors, especially Hamid Karimi (3 Com), Trevor Dearing (Bay Networks), Chris Gabriel (Cabletron), Bill Erdman (Cisco Systems), Ian Cowburn (Digital Equipment), Geoff Bennett (FORE Systems), and Michael Szabados and Mark Powell (Newbridge Networks when it was still UB Networks) for information about their companies' products and about standards and virtual LANs in general.

NOTE

I have been asked to remind all readers that information about the standards, particularly those from the IETF, is freely available on the Internet, as is much information about the vendors and their products.

CHAPTER 1

Introduction

Before considering the whats and hows of virtual LANS, it is necessary to understand the changing nature of organizational structure and the ways we do business today. (This is explored in Chap. 2.) It is also vital to be fully versed in the structure of networks, especially switched networks. To make sure that we—that is, I as author and you as reader—are all working from the same perspective, an overview of networking is given in Chap. 3. From there we can move on to consider virtual LANs (from now on called VLANs): what they are, and how they work.

What is a virtual LAN? With the multitude of vendor-specific VLAN solutions and implementation strategies, defining precisely what VLANs are has become a contentious issue. VLANs can be seen as analogous to a group of end stations, perhaps on multiple physical LAN segments, that are not constrained by their physical location and can communicate as if they were on a common LAN. A VLAN is basically a limited broadcast domain, meaning that all members of a VLAN receive every broadcast packet sent by members of the same VLAN but not packets sent by members of a different VLAN. All the members of a VLAN are grouped logically into the same broadcast domain independent of their physical location. Adds, moves, and changes are achieved via software within a VLAN. No routing is required among members of a VLAN. The different types of VLANs possible (and there are several) and how they work are described in Chap. 4.

Standards

With so many different VLAN ideas and implementations around, it is essential that a standard be set up. The body that has taken on the task is the IEEE (Institute of Electrical and Electronics Engineers), holder of many of the LAN standards. However, being LAN-oriented, this ignores the work of the other standards bodies in the area, namely, the ATM Forum, the IETF [Internet Engineering Task Force; best known for IP (Internet Protocol) internetworking standards], and the NMForum, for network management—even though it could be argued that VLANs are a management matter. This has resulted in a proliferation of standards around the edges of the topic, which all bear on each other, and have far-reaching effects on the development of VLANs overall. So far, two VLAN standards have been proposed for use, one based on IEEE 802.10 and the other on IEEE 802.1Q A standard is already set for Emulated LANs under ATM, by the ATM Forum. All the relevant standards are described individually and summarized in Chap. 5; meanwhile, the idea of standards in general is worth considering.

Before any standard for VLANs became moot, some vendors worked out their own VLANs—some purely proprietary, others based on an IEEE standard, but intended for another purpose: security tagging. ATM vendors ignored the issue because it was assumed that when LANE (LAN Emulation) came along, it would take care of it. The IETF had other things on its collective mind, such as inventing dynamic IP allocation. Unfortunately, it turned out that some VLAN memberships are based on IP addresses, and interoperation with the new and very useful IETF protocol would make the VLANs difficult and, in some cases, impossible, to operate in this way. These examples (and others could be cited, but the point is made) show that just saying "there is a standard for VLANs," which at the time of writing is expected to become possible by late 1997, is not the end of the problem. The next set of problems then appear:

- Migrating proprietary VLANs to the new standard
- Migrating security-tagged VLANs to the same standard as everybody else
- Adding to existing VLANs so that they can use sophisticated internetworking standards

In addition to all this, it is necessary that the vendors, their customers, and the standards bodies decide on what is actually (first) needed and (second) wanted out of the idea of virtual LANs—what are we all working toward. At present, some are simply additions to the network that make it work more efficiently, others are ambitious schemes that make all devices work seamlessly together, and still others are a reinvention of the business practices possible.

Network Management

Network management is discussed in Chap. 6 in terms of relevant SNMP (Simple Network Management Protocol) and RMON (Remote Network Monitoring) functionality and standards. After all, organizations which have installed high-speed and switched networks and advanced technologies, such as VLANs, do so in order to improve network performance. Two of the network manager's most popular management allies have been the protocol analyzer and the SNMP manager. The protocol analyzer allows the network manager to instantly go into a problem network and find and fix the problem which causes a failed or dysfunctional LAN. For many years, another popular tool used to "listen" for a properly functioning network has been the SNMP management station. (This "listening" function is explained in Chap. 3, section on carrier sense multiple access.) If the network is functioning properly, it is the SNMP management station's primary task to display and log these network statistics in easy-to-interpret information management displays. RMON, for network monitoring, is a widely adopted industry standard for the retrieval of network statistics from remote devices. It comprises two elements: an agent and a client. The agent builds up information within 10 RMON groups. By combining RMON and network management tools with distributed data collection and analysis consoles, a detailed picture of traffic flow patterns, network utilization, and protocol turnaround times can be constructed.

Security

Networks are valuable, both as capital investments and as a critical part of the enterprise. Valuable assets need protection, and corporate managers must protect these assets. The security of the network infrastructure itself, then, is an issue. The management and self-management of the network must be protected, or the network can be deliberately or accidentally brought down. Unfortunately, VLANs do little to ease the security burden, but may actually add to it. Chapter 7 discusses the need for security, how to improve it, and how to ameliorate deficiencies in current VLAN structure. For example, if security is paramount, then the cruder, simpler, types of VLAN are better, although they increase the network manager's burden. Policy-based VLANs, although they sound as if they could implement security policies already in place, are among the least secure. With remote access to a LAN where a VLAN is in place, security is obviously of concern. Chapter 7 also goes into the mechanisms of implementing security policies within VLANs and ELANs (Emulated LANs), the standards, devices, and protocols that protect our networks.

VLANs—Benefits and Problems

There are good reasons why so much attention is being paid to VLANs right now; with large networks being converted to a switched structure, it makes sense to implement VLANs—making the network more efficient and obviating the problems that can come from having a fully switched network, given that switches are, in essence, simply multiport bridges. Costs can also be reduced. Traffic between VLANs is firewalled.

This limits the propagation of multicast and broadcast traffic between VLANs. There is a consistent representation of a VLAN across a VLAN fabric (including one across ATM), so that the shared VLAN knowledge of a particular packet remains the same as the packet travels from one point to another in the VLAN fabric. Chapter 8 looks into these benefits as well as problems not only with implementing VLANs but also the difficulties they can themselves cause. For example, there can be interoperability problems and increased management complexity. It is possible for a VLAN to degrade, rather than enhance, performance. It can also be costly to install and can cause premature discarding of otherwise good equipment. Costs are not only capital—they can also be a matter of time, and some VLANs can be costly in terms of a network manager's time. Others instead can be inflexible from being too automated.

The Vendors

Given the diversity of possible VLAN installations at present, it is wise to consider the major offerings, each on their own merits. In Chap. 9, therefore, the vendor scene is described. There are five large companies in networking: 3Com, Cisco, Bay Networks, Cabletron, and Newbridge Networks. There are lesser vendors which also cannot be ignored, often because of their size in a different market, or because of their large installed base. These vendors have often arisen from a base of competence in one area only to, nowadays, encompass all or most networking technologies. There are other companies which specialize in switching products, and which have made a name for themselves in that niche. Their products are examined with particular relevance to VLANs, and the major offerings are examined in detail together with a few hints on purchasing strategy.

2

Virtual Organizations*

Businesses, along with the rest of society, are rapidly changing. Many people have told us that society is changing both its private and public faces. Who can deny it when we can see families around us breaking, remarking, and blending, or when we see millions of unemployed people (and usually not from choice) while most of us enjoy longer breaks from, and shorter days at, work?

---

*This is not an academic text, and I do not want to irritate the reader with constant footnotes and references. For more details on the subject matter of this chapter, the reader can refer to the publications listed in the Bibliography at the end of this book.

There are three basic types of trades:

- *Agriculture and mining*—growing on and extracting from the land around us
- *Industry*—manufacturing from raw materials
- *Services*—including health, education, and policing and other state services, but also private-enterprise services such as banking, retailing, and hairdressing

Today we must add a new type: information. This has to be separate from the others. If it is software, it is partly manufacturing—not from raw materials but from brain power; this means that it can be produced once and sold many times, except for the small cost of the delivery media, manuals, etc. It is often broader information—reports, research, project management—anything which is dreamed up by active and creative brains. It has been said that we have entered a postindustrial age: the Information Age.

Organizational Trends

The nature of work and the part it plays in our lives is also shifting; we and our employing companies are becoming more flexible. This flexibility takes several forms, including

- Who is employed
- When we work
- How we work
- What we do
- Where we work
- The tools we use
- The workplace

If the nature of the organization is changing, then our IT infrastructures need to change, too, or the dying mainframe syndrome will overtake us again in new fields. Let us therefore take a closer look at each of these categories.

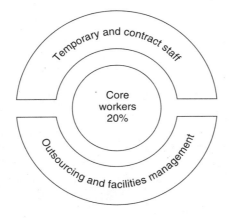

Figure 2-1
The makeup of the workforce.

Who Is Employed

Organizational structure has always relied on nonemployees for some of the work (see Fig. 2-1). While there is a core of full-time employed people, there are suppliers of goods and services, contractors, part-timers, teleworkers, and temporary help. This may have always been so, but the balance has shifted from the traditional 80 percent of core, full-time employees, with 20 percent coming from outside the organization's actual building to more like the reverse. We treat the different types of people differently, however.

Types of Workers

Personnel have always been divided into managers and staff—this has not changed, but the ratio has. Dogma has it that a manager cannot manage efficiently more than five or six people, and at that, does nothing else. This has been overturned as costs have been cut, with much flatter structures resulting. There may be only four layers between a junior staff member and the board of directors even in a large company. In these terms, there are basically three different types of companies:

- Pyramid (i.e., hierarchical)
- Flat (no line management)
- Matrix (where teamwork and cross-reporting are the norm)

While a pyramid company was the usual structure in any large enterprise until 20 years ago, this is now seen as wasteful and old-fashioned. Manufacturing, heavily influenced by the success of the Japanese model, has largely gone over to a flatter structure, often team-based. Office culture has accepted the flat model, often with matrix management. There is mixed success—anyone who has experienced cross-reporting will know only too well the difficulties of getting conflicting ideas and orders. With good communications, however, this can work well.

Outsourcing, Outtasking, and Facilities Management

- *Outsourcing* is the wholesale export of a whole process to an outside company, such as parts manufacture.
- *Outtasking* is the export of a discrete task, or series of tasks, to an outside agency, such as the preparation of a white paper.
- *Facilities management* is the contracting of an outside agency to do anything from electrical maintenance to cleaning the building—but always on the company's premises.

Manufacturers have bought in ready-made or part-assembled goods—this is a form of outsourcing, even if not usually thought of in that way. In Japan, most of the automobile industry grew from a quasi—cottage industry of small workshops supplying the larger factories for later assembly. In IT worldwide, many companies supply "badged" devices for other vendors on an OEM (original equipment manufacturer) basis. This is often to take advantage of lower labor costs in different parts of the world, but is also often simply to remove some of the manufacturing responsibilities and risks.

Manufacturing worldwide has also moved nearer to just-in-time (JIT) supply. This puts the pressure on distributors as well as manufacturers to keep stockholding costs down while still being able to supply any goods that are needed. Distributors and resellers therefore become partners in the business, with a lot of customer satisfaction dependent on them. They often also have responsibility for a lot of the customer support and service.

Services, too, have been supplied for many years. How many companies clean their own floors, do their own laundry, or employ their own cooks nowadays? Very few. Laying cable, creating networking infrastructures, and supplying customers with technical services has formed the basis of wealth for a whole army of new companies since the 1980s, and has been the saving of several large, older, IT companies.

Companies cannot afford to keep a full complement of workers sitting with nothing to do for some of the time. They nowadays bring in contractors for fixed terms to work on particular projects. A register of such contractors will be drawn up, and these people will become familiar friends in the company. Sometimes these will be women who need to devote a lot of time to family and cannot work all the time, sometimes retired people, often ex-employees, but often these will be full-time, professional consultants who operate a patchwork of contracts during a year for several companies, often, in fact, competing companies—these people have to be trusted.

While contractors are often well paid to compensate for their lack of security, part-timers usually get the short straw, with low pay and poorer, and less regulated working conditions. This will change as these days part-timers are more often expert workers, such as women who have returned after maternity leave or job-sharers. No company can afford to treat such people badly, despite any perceived lack of commitment to the corporation (often cited as an example of why part-timers are treated in a more offhand manner than full-timers). Contractors and freelancers who come and go at need are increasingly used and better organized (in that they are working this way nowadays often from choice, not necessity, and also that there are organizational tools, such as directories to find them) now that this is an accepted part of the workforce.

Teleworkers, often also part-timers, are increasingly common, especially if you include all types:

- *Road warriors*—usually salespeople, with laptops
- *Home workers*—sometimes or always work at home
- *Telecottagers*—work from a remote independent office near their homes

It is necessary to distinguish between those who are home-based or on the road and come into the office seldom and those who escape from the office on a frequent basis, or who work after hours at home, but who are officially office-bound. The former do not have a desk or dedicated computer at the office. The latter do, and for work done at home often use their own computers. The true home worker raises problems of IT support; both bring new security headaches to the enterprise such as outsiders viewing company data, viruses, and uncontrolled use of (often pirated) company software. Telecottagers bring a different kind of security risk—they will often share this telecottage with similar workers from other companies.

Temporary help has in the past been seen as mainly secretarial, but now this can be anybody brought in to take the strain while someone is on holiday or maternity leave or sick, or just for a short-term project. There is obviously going to be a lot more overlap with the contractors than in previous times as temporary staff are used for increasingly professional work. The main difference is in the type of contract. Instead of working on a fixed-term basis, temps are often hired and paid on a weekly or, more rarely, monthly basis.

When We Work

This can be divided into how long we work altogether, how long we work in any one week, and our hours of work.

It has been worked out that in one generation, we have cut our total lifelong working hours from 100,000 to 50,000 in one generation, at least in the developed world. This has come about from a mixture of longer education leading to a start of working life in our mid-20s, earlier retirement (note the increasing selling into the "third age" market as a result), longer holidays (especially in parts of Europe, where 30 days' paid leave is common), and a shorter working week.

This leads to the second part: hours of work. The working week in Europe is now often as low as 35 h per week for professionals (a little longer in most parts of the world). For many such people, paid overtime is not on the agenda, and the actual hours worked can be far longer. In this time of change, we are in a sort of interregnum, and this practice can lead to some people being severely overworked and others denied jobs altogether. The phrase "cash-rich, time-poor" has been coined recently to describe such overworked, underpaid professionals, who can be physicians, financiers, information workers, or network managers, to name only a few. Even so, most people today work fewer official hours than did previous generations, going from—in manual labor hours— about 45 h per week 30 years ago to about 40 today to—for white-color workers of all kinds—from about 40 hours per week 30 years ago to about 37.5 now. All these hours vary with geography, but the downward trend is there almost everywhere.

In many parts of the world, especially Europe, staggered-hour schemes (such as flexitime) have been introduced, originally to cut traffic problems, but are now increasingly enthusiastically adopted by employees.

How We Work

For those not on such schemes, even the times we work have changed because the expectations have changed. No longer clock-watchers, we are asked to "do the job, not the hours" and be flexible about it (see the preceding paragraphs). This can have its downside, as explained above, but also its upside—if we have done what we needed to do, we can leave early to attend to family business or take a long lunch to get to the supermarket. Company bosses like the idea of getting the job done, but the attitudes of many managers have not yet fully changed. People still leave the office early and so on, but often this is when they take matters into their own hands, and do it without anyone noticing.

In these changing times, very often the workforce is more advanced than their managers in these terms. They embrace—even sometimes push through—new working practices, sometimes against strong resistance from managers. Quite often a new directive comes from the top of the company and is adopted with pleasure by the workforce in general, but the middle managers are not certain about the benefits and worried that they have no metrics to ensure that productivity is kept up, and resist. Such companies have only themselves to blame—any new scheme coming in should be well organized. For example, it is often the case that the CEO (chief executive officer) says that working from home is to be encouraged to comply with local pollution laws. Workers begin to do so, but the middle manager is worried that without supervision, the employees will not be spending time working, but doing other things. And they may well be—but if the work gets done, and usually productivity actually rises, then, what does it matter?

Another new idea is *empowerment*. Again, the idea often comes from the top, but if workers attempt to implement it without managerial cooperation, they will find that they are not given budgets to manage their own workload and cannot hire and fire as they wish, leading to the new idea sinking without trace and a lowering of morale.

Similarly with teams. Everybody supports the idea of teamwork, and being perceived as not being a "team player" is a sure-fire way of not getting a job today. It has been shown that creativity and productivity rises when people work in self-contained teams that then cooperate with other teams. This has been practiced everywhere from units within offices to units in factories producing heavy machinery. Empowerment comes along here, too, of course, to make a team truly self-sufficient, but, again, is often a sham.

Still, with the new ideas around, the workforce often relies on itself and peer networking ensues. This is not in a formal, unionized way, but is a loose structure of help, support, and mentoring. Quite often today, this extends into networking with those in other companies doing the same job, making a mockery of competitiveness at one level. Again, this raises security risks. It also means that a workforce already more mobile because of the higher education that led to paper qualifications (which are more transferable) finds it easier to move on. A corporation today that adopts new ways of working has to reward its people and work to keep morale high to combat this. A corporation that tightens up in the face of perceived looseness in working practices, however, is doomed to a gradual decline. The world of work is moving on.

What We Do

Part of being flexible is to "do whatever it takes," as the corporate saying goes. With fewer staff doing more work (another corporate slogan is "work smarter, not harder," everybody has to be multiskilled. There are no secretaries to type letters except for top bosses, and they have to be highly adaptive administrators rather than just secretaries. There are not enough technical staff to sort out problems with every computer in the company—everybody has to be able to fix minor problems, or has to know someone who can. Again, the peer-level networking helps out here. Everyone, whether dealing with customers or not, must have an attitude of customer care, even where the customers are internal (within the company). In any project, there must be an awareness that if a process is held up, this can have an immense knock-on effect, given that every company works with very tight margins in terms of time and money.

Where We Work

Businesses operating internationally are using their staff more effectively by operating on a global basis. Such companies must encourage their staff to adapt to cultural differences, which can be as wide as the oceans separating the offices, even between England and the United States.

Again, the ideas may be supported at top and bottom, but it is middle managers who have to hold to the new standards and ethics—what is kidding along in one country is sexual harassment in another, for example, and they may find it impossible to understand, let alone enforce, such standards. A multinational culture is constantly adjusting with input from its workforce.

A global business needs more complex networking as well as culture than when business is kept within one country, as mail and groupware need to traverse the wide area across global pipes—woe betide the company with a misconfigured router—but it also means that everybody in the enterprise has to be aware of the need to communicate effectively with email (electronic mail), groupware, teleconferencing, and, now, videoconferencing. All need to be able to cope with many time zones as well, meaning that by combination of these factors, answers to questions may not arrive until the next day, a bottleneck which may cause havoc in a JIT microeconomy. Anyone working in this way needs to possess good time-organizing ability alongside other skills. However, at least this means that everybody gets used to being remote from someone they need to talk to, wherever their location.

Working from home, at least part of the time, is now commonplace, and full-scale telecommuting (see preceding discussion) is becoming more frequently used, especially in the United States, where this is the route often used to cutting traffic problems. We also work from our cars, with computers, phones, fax (facsimile) machines, and modems to keep us working. Similarly, we sometimes work from hotel rooms and hotel business centers. We send out consultants, technical support people, engineers, and even physicians into our customers' premises—we still expect them to be able to at least receive their email, and increasingly, to be able to get at their data in their own LANs.

Within the office, where our workspace has changed from separate, small rooms (where the fewer the people in any one room, the higher the person's status) to a largely open-plan environment. This varies; in some companies, nobody, not even the chairperson, has a separate office, while in others there is a half-way state, where managers have offices, but most people have desks in the open or are separated only by screens or in cubicles. There are usually rooms set aside for internal meetings to avoid wholesale disturbance. Given the flatter structure of organizations today, few managers need offices in this type of setup. With the emphasis on teams and empowerment, managers tend to take a more back-seat role than formerly; one trivial result is the lack of nameplates on doors. This latter is also useful because, as mentioned, we are all expected to be flexible—and that means constant rethinking of spatial needs. Groups and teams often move and re-form, needing different areas. Like supermarkets, where sales rise with every stock reorganization, a state of constant change is expected to refresh the workforce.

The Tools We Use

People who always worked remotely from the office (e.g., salespeople) are now better equipped to be completely mobile, allowing them to be even less centrally based. Computers not only are smaller than ever but also contain built-in fax/modems, making them more complex, too. There are also portable printers. Mobile phones are becoming more sophisticated as they get smaller in size. Pagers keep us in touch.

Some tools are software; for instance, network managers who need to troubleshoot remote parts of their networks can monitor and manage them from their own offices, can configure routers, and can update software, without a plane ticket or even a screwdriver.

The tools within the office are changing, too. Installations have to be more flexible in their arrangements, so that people can move around from place to place, from project to project, without creating havoc. Thus, many offices have only portable computers for their staff. A large network may have directory services, so that anyone can work from anywhere within it and still have the same rights and views. Many manufacturing premises have changed in a different way to paperwork-based offices, and they are still changing. The tools may include robots, depending on industry, which need far fewer staff to run than do older, labor-intensive factories. The term *teamwork* now commonly means that everybody in a team needs to be able to use the computer and other high-tech(nology) tools as they switch around flexibly, as in other places.

The Workplace

The office itself is changing. When so many people are working away from the central base for whatever reason, it is wasteful to allocate so much space to a desk per worker, and we are beginning to see "hot-desking," in which those who are in the office only occasionally find whatever space they can in an ad hoc way. Some companies, among them Digital Equipment, Sun, and Xerox, already use hot-desking as a way of life and nobody has an assigned desk—most or all workers in a unit or in a company keep all their possessions and work in progress in "carts" (U.S. term) or "trollies" (U.K. term), which they use when present in the office. This saves capital costs (smaller buildings) and utilities bills, too (less area to heat).

As mentioned above, the workplace has become more open, with fewer closed spaces. Such doors as exist remain open unless for a reason, to emphasize accessibility. With such open spaces, flexibility is better enabled for formation and re-formation of project teams.

Implications of Changes for IT

It is not only large enterprises that employ high-tech equipment today. The networking vendors are making a huge effort to capture sales in what they perceive to be the newest market in the developed world: the very small business. 3Com, for example, defines this as less than 20 users for its "tiny LAN" products. Of course, this could be a remote office within a large corporation, but the thrust of the marketing is aimed at the standalone small business, which has IT needs like any other, but probably no overall IT strategy and certainly less sophisticated technical support capabilities. As we become an increasingly information-based society, where bits of data are traded as much as atoms of matter, all businesses, including the local plumber or hairdresser, need the benefits of computers and communications. Sheer speed of data transfer has shot up over the last 10 years. In the 1980s, fax to some extent replaced the postal service; in the 1990s, email is becoming pervasive, replacing more post and fax. Cost and convenience may have been the stated reason, but a further result is immediacy. With email, you can write a message and get an answer within minutes. The time saving over post or fax is that no one has to deliver or transmit it; it has been shown that email is read on a higher-priority basis than either post or faxes, as it is easy to hit the REPLY icon and send it right back.

Technology and economics are a potent harbinger of change. This change has resulted in the possibility of more interesting jobs for all, including—as was seldom the case far in the past—women as well as those who can cope with a change in the way we work—in teams, with consensus rather than by direct orders, and when and where we like. IT structures in corporations must be able to cope, too—with more demanding workers, more complex networks, and ever-larger (higher-functionality) software applications.

Project teams form and change over time to meet evolving business conditions. As project teams change, their computing and networking environment must adapt. This often places a heavy burden on the network manager to continually reconfigure the network and reallocate network resources. What is needed is a flexible network that automatically adapts to changing conditions.

Flexible workers on variable-hour shifts work at odd times. No longer can the network manager back up the network happily at 6 P.M. knowing that no one will mind slow access times. The system must now be automated for the middle of the night, and even then there may be someone in a different time zone trying to get in; or it must be continuous, calling for more resources to be spent on this function. The main implications are, however, for security, support, and access needs.

Security

Having remote workers is a problem as well as a solution—along with the flexibility comes security problems. No longer can security cordons be drawn tight around a workplace LAN and be ignored thereafter. As soon as the wide area is accessible from the local area, there is always a worry that prying fingers are going to hack in. In fact, such security breaks from outside are usually trivial; most of the damage is done through internal negligence or malice.

Apart from the risks from remote workers allowing, unwittingly or otherwise, other eyes to see corporate data, independent technicians local to the remote site may be called in to sort out desktop devices—these are experts who can extract data, if so-minded, without attracting attention to the fact.

As mentioned earlier, contractors need to be trusted friends of the company, and need access to certain parts of the company's data, whether they are based inside or outside premises. This does not mean that they should have a free hand, especially considering that next week they may be at a competitor company doing similar work.

Today we have intranets and extranets. The former is a Web-enabled mass corporationwide sharing of data; in the latter case intranets from different companies are allowed a limited view of each other. An example is the networking company Cisco—everybody can view its Web page, but only authorized viewers (such as distributors) can access specific parts of the page. Those authorized viewers are doing so, very often, from within their own intranets.

To ensure security, particular attention needs to be paid to the question of whether to route or switch. While many networks now are based on a switched infrastructure rather than routing between domains, there must be routers, with their superior security functionality, rather than switches to keep the WAN (wide area network) at bay. Even within a network, in certain areas it may be preferable to keep a router. Switches often have routing built into them, and no doubt this technology will become more sophisticated, but it is probable that routers will remain as the main focus and switches will not match them for some time. (See Chap. 3 for descriptions of these devices.)

Support

Remote troubleshooting, configuration, monitoring, and upgrades are now possible from central sites, saving costs in terms of time and money for the network management staff, but complexity means that they have no more time for support than formerly. With the technical personnel so busy, there is little time for any desktops to be attended to except in emergencies. This means, as mentioned earlier, that users have to help themselves and each other. Trouble spots are often PC/Windows malfunctions, hard-disk crashes, and email failure. Some of these functions will be down to individual machines; others will be networked applications or hardware. Here, flexibility helps out—when the network is down, a lot of workers go home to carry on with other work. Previously, everybody tackled the filing.

Remote workers are often reduced to bringing in dysfunctional machines and doing without meanwhile, which is not very productive. Where remote support is impossible, independent technicians local to the remote site may be called in, as mentioned above. Otherwise, centrally operated management software (as described above) can be used, but this will not help in the case of simple, unmanaged devices such as modems, faxes, or printers.

Access Needs

To implement flexibility, companies need to have a flexible network. This means a switched network in many cases, with sophisticated software and hardware. Directory services and groupware enable everybody to see what others are doing and where they are. Remote personnel can then get at all the data without requesting help from anyone in the office via modem calls to the office direct or, increasingly, by tunneling through the Internet.

Some people will have laptop computers that travel with them; home workers will often have their main machine at home. Part-time home workers will often have their own computer; sometimes the company gives out obsolete models for such use. Remote workers all need to have either modems or ISDN (Integrated Services Digital Network) terminal adapters or routers. The remote communicator has been part of a proprietary system in the past, but increasingly, open, standard, commodity pieces of hardware and software are replacing the holistic systems. Such volume sales as this implies are a sign that remote access is no longer the prerogative of the few but is in the realm of the many.

It is imperative that all workers can access the data they need. This means that wherever they are, they must not be held up by not being in a certain broadcast group; they must not be locked out. This is where the subject of this book comes in—virtual LANs can circumvent this problem, and do more besides encourage the much-vaunted flexibility. With the new organizational structures and ways of working, here is one new tool that can bring order out of the chaos that networks today can be.

Network Construction Basics*

The network is best considered as a corporate utility rather than a collection of devices. As applications such as groupware, imaging, and client/server database begin to push bandwidth requirements to their limit and users look to minimize response times, higher-speed technologies—FDDI (fiber distributed data interface), 100Base-T, ATM (asynchronous transfer mode)—and switched networking will gain ground. Perhaps the most significant trend is toward switching in the LAN, whether Ethernet or ATM switching, at all levels of the network, from the workgroup to the backbone together with the movement of the router from the center of the network to the periphery in response. It is into this scenario that the introduction of VLANs is making an impact—as a way of rationalizing the sprawling corporate networks that have now emerged.

---

*Parts of this chapter are reproduced with permission from various reports from Datapro Information Services Group. Please see Bibliography for further details (references by Ford, Mann, Raymond and Minoli, and M. Smith and Datapro *Glossary*).

At its most basic level, a *network* is a collection of hardware and software which connects individual servers, PCs, and workstations together for resource sharing, messaging services, and workgroup computing. A grasp of a few key ideas and a knowledge of the hardware and software that embodies those ideas are required to understand virtual LANs.

OSI Reference Model

The basic standard underlying much of the datacommunications technology today is the International Organization for Standardization (ISO) Open Systems Interconnection (OSI) seven-layer reference model. The ISO is a worldwide federation of national standards bodies from some 100 countries, one from each country (the national body "most representative of standardization in its country" is the phrase used when accepting members) and is a nongovernmental organization that was established in 1947. Its mission is to promote the development of standardization and related activities in the world to facilitate the international exchange of goods and services and develop cooperation in the spheres of intellectual, scientific, technological, and economic activity. An understanding of the OSI reference model is essential to an understanding of virtual LANs.

Layering is a basic structuring technique used in the OSI model. Each layer is composed of an ordered set of subsystems (even formal sublayers), with logically related functions grouped together. The OSI model breaks down internetworking activities between systems into four distinct groups of functions:

- Communications-oriented
- User-oriented
- Information moving (across a network)
- Information handling and formatting

There are seven layers of the OSI model that communicate between one end system and another end system (see Fig. 3-1). The layers cover nearly all aspects of information flow, from applications-related services provided at the application layer (layer 7) to the connection of devices to the communications medium at the physical layer (layer 1).

Below the physical layer, the medium itself corresponding to "layer 0"—such as wire, cable, or through-the-air communication—is not

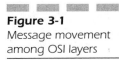

Figure 3-1

Message movement
among OSI layers

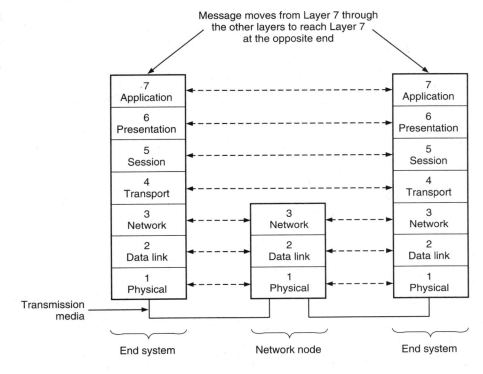

addressed by the model. Application, presentation, session, transport, network, data-link, and physical layers have been defined (see Table 3-1). The model described is OSI's seven-layer hierarchy with the purposes of each layer.

The seven layers can be divided into two functional groups: the transport platform (layers 1 to 4) and the application platform (layers 5 to 7) (see Fig. 3-2). The transport platform's function is to get data from one system to another without errors. The application platform's function is to interpret the data stream and present it to the user in a usable form.

Each layer contributes functions to the communications task. For example, the data-link layer enables communications across a single physical connection, while the network layer provides end-to-end routing and data relay. Services at the upper-layer interface—providing communications to the next-higher layer—are provided by each layer, usually described by a service specification for the layer. Services at each layer are provided by a layer entity. Each layer entity communicates with its peer at the same layer on another system, providing services specified in the service specification.

TABLE 3-1

The Seven Layers
of OSI

Layer	Name	Purpose
7	Application	Applications and application interfaces for OSI networks; provides access to lower-layer functions and services
6	Presentation	Negotiates syntactic representation for the presentation layer and performs data transformations
5	Session	Coordinates connection and interaction between applications; establishes a dialog, manages and synchronizes the direction of data flow
4	Transport	Ensures end-to-end data transfer between applications, data integrity, and service quality; assembles data packets for routing by layer 3
3	Network	Routes and relays data units among network nodes
2	Data link	Transfers data units from one network node to another over a transmission circuit; ensures data integrity between nodes
1	Physical	Delimits and encodes the bits onto the physical medium

Layers are sometimes divided into sublayers, for several reasons. Layer functions are often divided into separate modules to handle the service interface of the layer beneath it. This avoids "rewriting" the entire layer. For example, the link layer of the IEEE 802 Local Area Network (LAN) standards is divided into a logical link control (LLC) sublayer and a

Figure 3-2

Application and
transport divisions

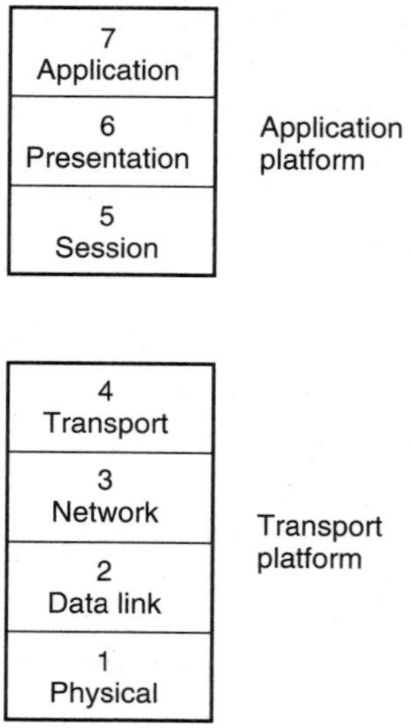

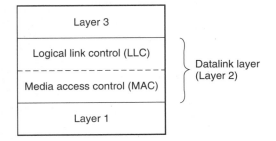

Figure 3-3
The data-link layer

media access control (MAC) sublayer (see Fig. 3-3). The MAC sublayer depends on characteristics of the underlying physical layer. Any layer may originate a message to fulfill its responsibilities. The message may not bypass any layer en route to its destination. If a message leaves the node, it will end up in another node at the same layer that originated the message.

There are good reasons for layering; it simplifies change, and components inside a layer can be changed without affecting any other layers in that node. Layers are like structured programming—but for teleprocessing systems. Because there are rigid interfaces between levels, fewer people need to react to changes, allowing the changes to be implemented faster. There is no better way of achieving complex functions. Layering allows each network function to be made "transparent," unaware and independent of other functions at other layers, thus enabling any layer to be modified without changing the entire monolithic architecture.

Each layer may support one of several different protocols designed for specific network applications; the choice of a specific protocol is optional, allowing users to tailor networks to their own design. Each layer defines functions crucial to the communications process at that layer, independent of the other layers. However, a layer may perform functions hinging on functions performed in the layers immediately above or below. A layer can communicate with another device or network node only at its peer layer. Messages exchanged between peer layers are "enveloped" with messages from other layers and passed through these other layers on the way to their destination, picking up and then shedding these other protocol layers along the way. For example, if layer 7 at one end system must send a message to layer 7 at another end system, the message must travel down through six layers at its own end and then up through six layers at the other until it reaches layer 7 at its opposite (peer) layer.

Each network node (a network user, computer, terminal) is equipped with this layer mechanism. However, not all intermediate nodes need all seven layers. Network nodes, in particular, must only route and transmit data packets—functions at the bottom three layers of the OSI model. Layer 4 through 7 functions are not required and, therefore, not included in network node software. Data packets processed in these nodes reach only layer 3 and are then routed elsewhere. A node communicates with its peer in another node sending or receiving data. Data transfer is routed from layer 7 down to layer 1 at the transmitting node, then along the network to layer 1 at the receiving node, and finally from layer 1 up to layer 7. Peer layers communicate by the same method.

The message initiated at the application layer is passed from layer to layer, through the various OSI layers, encapsulating control information in the process. A fully encapsulated message enters the cable at layer 1. The procedure is reversed at the receiving end. Each item of control information is processed at its appropriate layer, and the message itself passes up to layer 7. Data transfer essentially is a packaging process at the transmitting node and an unpackaging process at the receiving node.

The Layers

A number of objectives were considered by the reference model's designers: to limit the number of layers to make the system engineering task of describing and integrating the layers as simple as possible, to create boundaries between layers at points where the description of services can be small and the number of interactions across each boundary is minimized, and to collect similar functions in the same layer. Any layer may originate a message to fulfill its responsibilities. The message may not bypass any layer en route to its destination. If a message leaves the node, it will end up in another node at the same layer that originated the message.

The seven layers can be divided into two functional groups: the transport platform (layers 1 to 4) and the application platform (layers 5 to 7). The transport platform's function is to get data from one system to another without errors. The application platform's function is to interpret the data stream and present it to the user in a usable form (see Fig. 3-4).

Each layer contributes functions to the communications task. For example, the data-link layer enables communications across a single physical connection, while the network layer provides end-to-end routing and data relay. Services at the upper-layer interface—providing communications to the next-higher layer—are provided by each layer, usually

Figure 3-4
Layer 7 and onward

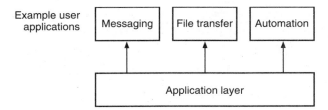

described by a service specification for the layer. Services at each layer are provided by a layer entity. Each layer entity communicates with its peer at the same layer on another system, providing services specified in the service specification.

Although not as widely known as the reference model, part 2 of that model (ISO 7498-2: Security Architecture) performs a similar function for security, formally defining security services and making recommendations on which services are appropriate to which layers. This is described in more detail in Chap. 7.

Here, I shall concentrate on layers 2 and 3, as these are central to switching and virtual LANs.

The Network Layer. The network layer (layer 3) provides the means to establish, maintain, and terminate connections between systems. Its basic service is transparent data transfer between transport entities.

The services provided by this layer encompass the following:

- Establishing network connections for transporting data between transport entities through network addresses
- Identifying connection endpoints
- Transferring network service data units
- Noting errors for reporting unrecoverable errors to the transport layer
- Sequencing network control data units
- Flow control
- Releasing the network connection

Routers use IP addresses to send data; this addressing scheme works at this layer.

The Data-Link Layer. The data-link layer (layer 2) provides the procedural and functional means to establish, maintain, and release data-link connections between two network nodes or network entities and to transfer data frames (or packets). This layer also detects and may correct errors that occur in the physical layer.

Services provided by the data-link layer to the network layer include data-link connection, sequencing, error notification, flow control, and data unit transfer.

Layers are sometimes divided into sublayers for several reasons. Layer functions are often divided into separate modules to handle the service interface of the layer beneath it, to avoid rewriting the entire layer. For example, the link layer of the IEEE 802 local area network (LAN) standards is divided into a *logical link control* (LLC) sublayer and a *media access control* (MAC) sublayer. The MAC sublayer depends on characteristics of the underlying physical layer.

It is the MAC sublayer of this layer that bridges and switches use to find the recipients of data. All devices on a network have a unique MAC address.

LAN Standards

The major standards-setting body in the LAN industry is the IEEE 802 committee. Formed in 1980 with the express task of defining industry-wide standards for LANs, the 802 committee has been one of the most successful independent standards-setting bodies in the history of the computer industry, and its contributions to advancing network interoperability and the general state of the LAN art have been immeasurable. The IEEE 802 subcommittees define standards for various segments of the industry. The 802.3 standard describes an Ethernet-like network; the 802.5 standard covers token ring; 802.2 describes logical link control—a method of interfacing network software to network hardware at the juncture between the physical and data-link layers of the OSI model. Both 802.3 and 802.5 networks implement 802.2 logical link control. The IEEE as a whole is the world's largest technical professional society. Founded in 1884 by a handful of practitioners of the new electrical engineering discipline, today's institute comprises more than 320,000 members who conduct and participate in its activities in 147 countries.

FDDI standards are set by ANSI (the American National Standards Institute). ANSI was founded in 1918 by five engineering societies and three government agencies, and remains a private, nonprofit membership organization supported by a diverse constituency of private- and public-sector organizations.

ATM standards are formulated by the ATM Forum and the ITU-T. The process adopted by the ATM Forum—based on simple majority voting—has allowed the standards-making process to progress rapidly, and with continued momentum. Nevertheless, much of the technology that the ATM Forum is using has originally been defined and developed via the ITU-T. The ATM Forum is a worldwide organization, aimed at promoting ATM within the industry and the end-user community. Formed in October 1991, with four (4) members, ATM Forum membership currently includes more than 750 companies representing all sectors of the communications and computer industries, as well as a number of government agencies, research organizations, and users. The ITU-T is part of the International Telecommunications Union (itself part of the United Nations), a worldwide organization within which governments and the private sector coordinate the establishment and operation of telecommunication networks and services; it is responsible for the regulation, standardization, coordination, and development of international telecommunications as well as the harmonization of national policies. It was set up in 1865—it is the most venerable of all standards organizations. Table 3-2 gives the major current LAN standards.

Ethernet and Fast Ethernet are mostly considered when changing to a virtual LAN, and we will look at these newer versions of Ethernet a little more closely. It is worthwhile to understand the IEEE 802.3 standards' shorthand notation. This is composed of three elements; for instance, for 10Base-T

- 10—megabits per second
- Base—baseband (or *broad* for broadband)
- T—twisted pair (or 5, for meters per segment divided by 106)

Under this scheme, it is easy to work out the newer names such as 100Base-TX, 100Base-FX, or 100Base-T4 for the three variants of Fast Ethernet.

100Base-TX.   Basically this is a renaming of the twisted-pair variant of 100Base-X. It borrows the physical characteristics of FDDI's TP-PMD (twisted-pair physical-media-dependent FDDI equivalent), but uses Ethernet framing and CSMA/CD (carrier sense multiple access with collision detection). A media-independent interface will allow a single interface card to use either this or the other flavors of 100Base-T.

100Base-FX.   This is really a renaming of the fiber variant of 100Base-X. Although it borrows the physical characteristics of FDDI's normal

Technology	Standards body	LAN type
Higher-layer interface	IEEE 802.1	Many subgroups for MAC control and management, including 802.1Q for virtual LANs
Cable specifications	IEEE 802.14	For all LANs
Logical link control	IEEE 802.2	For the control of the lower part of layer 2
CSMA/CD (Ethernet-type)	IEEE 802.3	CSMA/CD for Ethernet, Fast Ethernet, and switched versions; also Gigabit Ethernet (when not in full-duplex version) as 802.3z and full duplex flow control IEEE 802.3x
Token bus	IEEE 802.4	Token-passing bus LAN type, seldom used
Token ring	IEEE 802.5	For token ring, switched token ring, and duplex switched token ring
Metropolitan area network	IEEE 802.6	For distributed queue dual-bus subnet
Integrated voice and data LAN	IEEE 802.9	Still in progress
Standard for interoperable LAN security	IEEE 802.10	Includes packet tagging, sometimes used for virtual LANs
Wireless local area network	IEEE 802.11	Unfinished standard for wireless LANs which uses Ethernet bridges with roaming to join them to the network
100VG-AnyLAN	IEEE 802.12	100 VG-AnyLAN (VG—voice grade); runs at 100 Mbits/s, token-passing scheme
FDDI	ANSI X3T12	For FDDI and switched FDDI
TP-PMD	ANSI X3T12	For FDDI over copper and switched version
FDDI II	ANSI X3T12	FDDI II—not finished and not used much
Fiberchannel	ANSI X3T11	Used for Gigabit Ethernet; lower-speed (200-Mbit/s) standard is ready, but higher-speeds standards not yet
B-ISDN	ITU-T	Basic standard is out
ATM LANE	ATM Forum	LAN Emulation is ready
MPOA	ATM Forum	Multiprotocol over ATM standard is not ready at the time of writing but due soon

fiber PMD, it uses Ethernet framing and CSMA/CD. Note that the physical media support distances up to 2 km, but in a CSMA/CD environment (i.e., any 802.3 except full-duplex links), there is a maximum limit of 412 m between on the size of the entire network. A media-independent interface will allow a single interface card to use either this or the other flavors of 100Base-F.

100Base-T4. This is a renaming of 4T+. It uses 8B6T (three-state: 8 bits encoded into 6 trits) encoding and 25-MHz clocking, and in addition to the two pairs traditionally used in the manner of 10Base-T, it also has two pairs used in bidirectional half-duplex fashion. Among other things, this means that this particular kind of Ethernet cannot be made full-duplex without the use of more pairs. A media-independent interface will allow a single interface card to use either this or the other flavors of 100Base-T.

There may also be 100Base-T8 or 100Base-T2 at some point.

Fast Ethernet packets are identical to 802.3 packets (with bit times at 10 times the speed), but the nature of CSMA/CD requires that the overall radius of the net be limited to one-tenth the size of 10-Mbit/s Ethernet. A typical maximal system would be hubs on a very short backbone (up to 5 m) with the (repeating) hubs supporting links up to 100 m. Single-hub networks allow up to 325 m (e.g., 225 m of fiber for one link and 100 m of twisted pair for any other link). Extension of the net beyond this would require a switch, router, or bridge. Fiber links employing the CSMA/CD but with no hubs can run 450 m, and full-duplex links (i.e., with CSMA/CD "disabled") can run 2 km. It is obvious that without (at minimum) switches, this technology will be limited to connecting a few offices to a server at most. For the construction of a VLAN, it is best to have widely used variants across the whole network; these would be 10Base-T Ethernet or 100Base-TX or 100Base-FX. ATM is also used, and mixed with Ethernet and Fast Ethernet, but at the present time in very few implementations. Where it is used, LANE is necessary.

LAN Types

Topology

Networking on a Bus. The earliest commercially viable network, Ethernet, used (and still uses) a *bus*—a single data path to which all workstations directly attach and on which all transmissions are available to every workstation. Only the workstation to which the transmission is addressed can actually read it, however. A bus cable must be terminated at both ends to present a specified impedance to the network workstations. Therein lies one of the major disadvantages of a bus network, since any break in the bus cable takes the entire network down. Another disadvantage of a bus network is that it cannot be switched—and therefore cannot run a VLAN scheme.

Physical Star and Logical Bus. A new version of Ethernet—10Base-T—appeared in the late 1980s. It uses unshielded twisted-pair wiring and is arranged in a hub- or switch-based star topology. Yet the network *logically* acts as a bus; that is to say, signals transmitted by any workstation are available on the network at all workstations. Only the station for which a transmission is meant can read it. The star has no single point of failure like the coaxial bus—a cable break on a star disables only the station(s) that the broken cable connects to the hub or switch. Another common Ethernet fault that is alleviated by the hub-based topology is "jabber." Sometimes a network adapter card will fail in such a way that causes it to begin flooding the network with meaningless data packets. While a jabbering node will disable an entire bus-based network, and can be extremely difficult to identify on a hub-based star, an intelligent hub can simply turn off the jabbering node(s). This topology can be part of a virtual LAN if it is a switched network.

When this type of network was first introduced, end nodes were grouped to share ports on a hub. With the advent of switches and the lowering of prices, "users each need 10 Mbits to the desktop" became a mantra repeated without thought. This is not true; any improvement on the old 10 Mbits/s shared between anything up to 50 users would show results. Now we hear the same cry, upgraded to take account of Fast Ethernet. It is true that having only one end station per port on a switch gives added advantages, but not limited to bandwidth. It is here that added management and the entry of VLANs can make such a difference to the enterprise network as a whole, as well as relieve bandwidth bottlenecks. While 100 Mbits/s each may in the future become necessary, at present this speed works very well—it is the switching that makes the difference.

Networking on a Ring. The simplest form of a ring topology is one in which each workstation is connected to another adjacent workstation in a closed loop. Such a ring, however, shares with bus-based Ethernet the disadvantage of a single point of failure. For this reason, all ring networks in use today are actually implemented as physical stars.

Physical Star and Logical Ring.   Both token ring and FDDI implement a ring topology on a physical star. At the center of the star is a device similar to a hub called a *multistation* (or *media*) *access unit* (MAU). Multiple-MAU networks actually form a ring, since the MAUs are connected via ports at either end of the devices. These ports are called "ring in" at one end and "ring out" at the other. In this case, however, if a cable break occurs between MAUs, the internal circuitry of the MAU can loop the ring back on itself to bypass the break and keep the ring functioning.

Access Method

How can only one computer at a time be allowed to transmit on the network? Access to the network—the right to transmit—can be allocated in two ways: randomly or in a deterministic order. In a random access method, any station can initiate a transmission at any time—unless another station is already transmitting. In a deterministic access method, each station must wait its turn to transmit. Only one example of each type of access method is widely used. Carrier sense multiple access (CSMA) is the random-access method used with Ethernet. Token passing is the deterministic method used in token ring and FDDI networking.

Carrier Sense Multiple Access.   Ethernet uses CSMA, a random-access method that requires a workstation to "listen" to the network media before attempting to initiate a transmission. If the workstation "senses" activity on the line, it defers its transmission for a short period before trying to access the network again. When it senses a clear line, it begins sending. A major shortcoming of this access method is that two workstations can sense an unused line and begin transmitting at approximately the same time. The result is a collision. Ethernet implements carrier sense multiple access with collision detection (CSMA/CD). A collision is "detected" when the voltage level on the network equals or exceeds the level generated by two stations sending at the same time. Any station can detect this condition, and when one does, it begins sending a jamming signal that forces each transmitting station to stop transmission. The transmitting stations then wait for a random amount of time before attempting to transmit again.

CSMA/CD and random-access methods in general work best when network traffic tends to be unpredictable and bursty, consisting of many short transmissions. Performance of a CSMA/CD network can degrade quickly under the kind of sustained heavy loads generated by many workstations sending large files. Paradoxically, Ethernet is still most widely used in just such an environment—connecting powerful UNIX-based engineering and graphics workstations.

Token Passing.   FDDI and token ring use *token passing,* a deterministic access method that allows only one station at a time the right to access the network. A special data structure called the *token* passes from station to station in sequence. A station that has data to transmit grabs the token and changes a bit, transforming the token into a packet header. When the data is received, the altered token is placed back on the ring as an acknowledgment from the intended recipient that the data was received without error. The transmitting station then generates a new token and passes it to the next station on the network.

Summary.   The basic defining concepts of networking are topology, media type, and access control. There are several different network types commonly used today; the most popular are Ethernet and token ring.

Ethernet/802.3 is a 10-Mbit/s bus-based network that can run on two types of coaxial cable, on unshielded twisted-pair telephone wire, or on fiberoptic media. It uses a random-access method: CSMA/CD. The 802.3i standard defines 10-Mbit/s Ethernet operation over fiberoptic cable, also known as 10Base-F. 10Base-FL (link) is an asynchronous fiber which supports 10 Mbits/s up to 2 km and uses a star topology. 10Base-FB (backbone) is a synchronous fiber and uses a star topology which supports 10 Mbits/s up to 2 km and significantly increases the number of repeaters allowed in series—it allows serial chains of repeaters. 10Base-FP (passive) uses a passive fiber star for a buslike performance. Ethernet is the most common LAN type installed today. Increasingly, it is switched, and often has uplinks to a Fast Ethernet or FDDI backbone. This is where we will find almost all the VLANs appearing. Ethernet runs on three types of media: thick coaxial cable, thin coaxial cable, and unshielded twisted pair (UTP). Only twisted pair is used in modern, switched networks. 10Base-T Ethernet runs on UTP and uses the RJ-45 modular connector. This can be category 3 or 5, but despite the extra cost, category 5 is increasingly used in order to get higher speeds and greater distances, and for "future proofing" purposes.

Fast Ethernet/802.3u is nearly identical to its 10-Mbit/s ancestor. The packet length, packet format, error control, and management information are identical to those of 10Base-T. It allows 100-Mbit/s transmissions over unshielded and shielded twisted-pair and fiberoptic cabling. It is implemented in a star topology and is often switched. Again, Fast Ethernet, especially in mixed Ethernet/Fast Ethernet networks, will host most of the virtual networks. Fast Ethernet also runs on unshielded twisted pair and uses the RJ-45 modular connector. Again, this can be category 3 or 5, but category 5 is almost exclusively used to gain reliability and greater distances than category 3 allows and is always used in new installations.

Gigabit Ethernet/802.3z combines fiber channel technology with Ethernet media access methods, running at 1000 Mbits/s (1 G-bit/s) over fiberoptic cabling (it has been tested with CAT 5, but not yet approved). The IEEE standard is expected to be finalized by 1998, although the first non-standards-based products have appeared in 1997 with standards-based products following later. Initial products require fiberoptic cabling, but work is progressing on developing a version to run over untwisted wire pair. Both 100VG-AnyLAN and 100Base-T media access methods have been proposed to run over the fiber channel PHY (physical) layer. Gigabit Ethernet is expected to run mainly on servers and in backbones on a single site (there are stringent length-of-cable limitations) and, again, will host VLANs.

Isochronous Ethernet/802.9 combines the data-handling features of Ethernet with the features of Integrated Services Digital Networking (ISDN) for voice and data. People don't really use the terms consistently, but what is usually meant is the distinction between data streams that can be synchronized to the network using a clock supplied by the data communications equipment (DCE) (synchronous) versus data streams that were generated by a clock that is independent of the network (isochronous). In the latter case, the DCE must use additional techniques, usually called *rate justification,* to decouple the data clock from the network clock. This usually entails a certain amount of additional overhead bandwidth.

100VG-AnyLAN/802.12 operates at the same speed as Fast Ethernet, but unlike Fast Ethernet, does not use the CSMA/CD media access scheme. Instead, it uses a *demand priority* scheme for time-sensitive traffic and appears to have technical advantages over Fast Ethernet. However, it lacks widespread support among the vendor community and switches for it are rare; the main vendor is Hewlett-Packard, which is not noted for its virtual LANs.

Token-ring/802.5 is available in versions that run at 4 or 16 Mbits/s and can operate on shielded or unshielded twisted-pair wire as well as on fiberoptic media. It is arranged in a star topology, with a hublike device (a MAU) at the center of the star. Its access method—token passing—makes the physical star arrangement act as if the stations were arranged in a ring, as the token is passed from one workstation to the next in sequence. To be part of a VLAN, a token ring will need to change to switched token ring, introduced by several vendors in 1996. Token ring runs on shielded twisted pair and unshielded twisted pair. The shielded twisted-pair version uses a 9-pin D connector at the card end and an IBM Cabling System connector at the MAU. The IBM connectors are called "hermaphroditic" because, unlike most cable connectors, they do not require male-to-female mated connections—any IBM connector can connect to any other IBM connector.

Arcnet is a proprietary token-passing network which is relatively cheap to install. Because of its early entry into the market, it has established a large installed base. However, it is rarely the network of choice today. The network is not standards-based and is slower than current standards, running at 2.5 Mbits/s. Arcnet continues because of its installed base, but will continue to decline as standards and speed dominate the marketplace. Arcnet can never be part of a virtual network.

FDDI (fiber distributed data interface) is a 100-Mbit/s network that operates over optical fiber, or, for limited distances, shielded or unshielded twisted-pair cable. It is implemented using a dual-ring topology—FDDI concentrators and "dual-attached" workstations require four separate fiberoptic connections—with receive and transmit sides for each ring. Dual attachment provides fault tolerance—in case of a cable break, an FDDI concentrator or dual-attached workstation can "wrap" the cable ends adjacent to the break onto the second ring. Single-attached workstations are also implemented. Single attachment allows many workstations to be attached to a concentrator in an economical manner. Since these workstations are not directly attached to the main ring, a cable break will disable only that workstation and not affect the entire ring. Single-attachment workstations can also attach directly to the ring. FDDI, in its original form, runs on fiberoptic cable. Two types of fiber media are in common use in the communications industry: single-mode and multimode fiber. Multimode fiber allows the use of light-emitting diodes (LEDs) as the light source and is capable of transmission speeds in the hundreds-of-megabits-per-second range. Single-mode fiber requires a laser as the light source and is generally used for long-distance communication, such as in the public telephone network. There is an equivalent of FDDI operating at 100 Mbits/s over UTP cable. This is known as *twisted-pair physical-media-dependent* (TP-PMD) and incorporates the old CDDI and SDDI standards.

ATM (asynchronous transfer mode) is the formal ITU-T standard supporting cell-based voice, data, video, and multimedia communication in a public network under broadband ISDN (BISDN). Cell relay is a high-bandwidth, low-delay switching and multiplexing technology. ATM allows bit-rate allocation on demand so that the bit rate can be selected for individual connections. In addition, the actual "channel mix" at the broadband interface can change dynamically. ATM supports channels with any fixed rate in the range of kilobits per second up to the total payload capacity of the interface. The ATM header contains the label, which comprises a virtual path identifier (VPI), a virtual channel identifier (VCI), and an error-detection field. ATM networks are designed to provide a fair allocation of bandwidth to users. The traffic profile and quality of service expected are parameters that are defined by users for the network. Part of this process of fair allocation of bandwidth is that users can change their transmission rates by reacting to feedback information from the network. ATM technology is designed to carry multiple media types generated by multiple application types. To do this ATM has to carry multiple traffic types. Today, this includes *constant bit rate* (CBR) (for video and voice applications) requiring constant-bit-rate circuits, *variable bit rate* (VBR) (for high-priority data applications such as inter-LAN traffic), and *available bit rate* (ABR) (which makes use of spare bandwidth that is not being used by CBR and VBR traffic). While the mechanism for ABR does not guarantee the availability of bandwidth, it does set out to provide fair access to the available bandwidth. For such a scheme to function, a flow-control feedback mechanism is required. This notifies end systems in the network that the network is becoming congested, allowing the end systems to react accordingly by reducing the ABR traffic entering the network.

LAN Emulation (LANE) has been designed to allow existing networked applications and network protocols to run over ATM backbone networks in a standardized way. It addresses the requirements of end systems which are directly attached to ATM networks, as well as addressing the situation in which end systems are connected via bridging devices (working at layer 2 of the OSI reference model). LANE allows multiple LANs to coexist on the same physically interconnected ATM network. LANE support includes Ethernet-Ethernet, Ethernet-ATM, token ring-token ring, token ring-ATM, and ATM-ATM and also allows for full-length FDDI frames to be transmitted across an ATM backbone without any fragmentation effects due to the LAN emulation. For further details of LANE, see Chap. 5. VLANs can span ATM networks by means of LANE.

Data Format

I have mentioned packets and cell throughout—it is time to talk more about the data types that are found in the different kinds of LANs. Packets arranged as frames are the data format used in all types of LAN except for native ATM LANs. A packet has a header that contains the address (in this case, being in a LAN, the MAC address of the station that will receive it) and a footer—usually a bit, or series of bits, to signify the end. The payload is the data. ATM LANs, although rare, do exist, and in these LANs there are no packets, only cells. A cell consists of an information field that is transported transparently by the network and a header containing routing information. In ATM, information is packed into fixed-size cells of 53 octets. Cells are identified and switched throughout the network by means of a label in the header. The payload comprises 48 octets.

The Role of Bridges, Routers, and Switches in Networks

Bridges

Bridges are used to connect two or more LAN segments together. They often include a filtering function which enables the bridge to send frames out only to the segment which contains the destination station. Bridges work at layer 2 of the OSI seven-layer model, making them protocol-independent. Bridges look at the MAC address to make forwarding decisions and do this without changing the Ethernet frame. Bridges have fallen from popularity as routers (with more functionality) have become cheaper and, more recently, as they are replaced by switches.

Routers

A *router* performs all the functions of a bridge and a lot more besides. Routers are used to improve network segmentation, and to route between dissimilar LANs, and also to route wide area connections.

Routers always use a store-and-forward technique. This means that the entire frame is read into memory before it is sent out. Routers read all

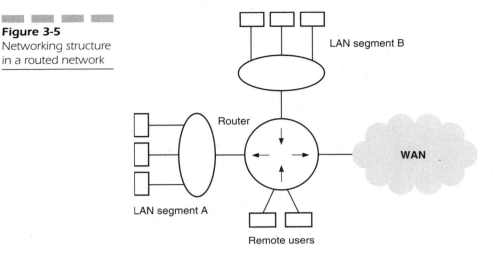

Figure 3-5
Networking structure
in a routed network

the layer 2 information and part of the layer 3 information. They are able to identify the sending and receiving stations by information in the layer 2 information. They can also identify the protocol in use.

This means that routers operate at layer 3 of the OSI seven-layer model, so they are protocol-dependent. A router strips out the layer 2 and address information and repackages the data into a new frame, which can then be transmitted across a different type of LAN segment.

Recently, the role of a router has changed. Instead of being at the center of a network with all traffic flowing through it (see Fig. 3-5), the bottleneck has been eliminated by the increased use of switches in this role, with the router relegated to the edge of the network and used to filter traffic to and from the WAN (see Fig. 3-6). Here it can provide firewall functionality without slowing down the whole network.

LAN Switches

Switches are basically multiport bridges, but share some characteristics with routers. Like routers, switches work by dividing up the network into a number of segments, each of which can operate without interference from traffic local to any of the other segments. Switching is performed at layer 2 of the seven-layer model—the same as bridging. Since it is performed at layer 2, the MAC address is used, which is independent of protocol address. Like a bridge, a switch learns which addresses reside

Figure 3-6
Networking structure
in a switched
network

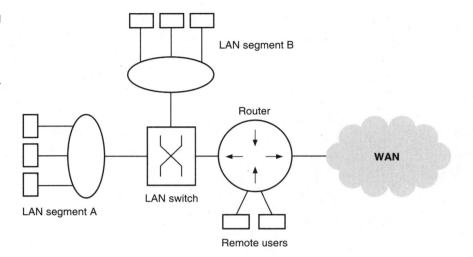

on each of its ports and then switches data appropriately. A switch can be designed using conventional microprocessors or dedicated ASIC (application-specific integrated circuit) technology.

Frame switches enable network managers to divide networks into virtual LANs (VLANs). Virtual LANs make changes easier and can improve performance, prevent broadcast storms, and improve security. Switching also introduces some new challenges, such as how to manage and monitor for problems in a network divided into many independent segments.

Switching Types.   There are five types of LAN switching:

Ethernet switching.   Based on the 802.3 protocol, but in the case of a fully switched network, need not employ the CSMA/CD protocol. Ethernet switches can run at 10 Mbits/s, 100 Mbits/s (then called *Fast Ethernet*), and soon, at 1 Gbits/s (then called *Gigabit Ethernet*). The bandwidth is used more efficiently than shared Ethernet as the data stream goes only to the user(s) it is directed to, rather than all of it running around the whole network.

Token-ring switching.   Token ring, unlike Ethernet, is deterministic. Therefore, the bandwidth is much more predictable. However, switching on a per port basis rather than sharing the ring will increase the bandwidth in a similar way to Ethernet switching.

FDDI switching.   FDDI switching is a fast standard (100 Mbits/s), but FDDI switches are rarely used in networks, since they are relatively expensive compared with Ethernet and Fast Ethernet switches.

Port switching. Port switching switches between backplane segments within a hub. For example, port 1 could connect to backplane 3 and port 2 could connect to backplane 1. The net effect is that several users end up sharing the same backplane, creating a "collapsed backbone," but offering more flexibility. Port switching is also known as *configuration switching*, and sometimes *desktop switching.*

Cell switching. Today, this always means ATM. Like frame switching, ATM is also a switching technology, but unlike frame switching, ATM switching—or cell switching—is connection-oriented, like the telephony network. In contrast, frame switching is connectionless—a packet can take any route across a network from its source to its destination. In cell switching, a connection is first negotiated and established across multiple switching points. An appropriate circuit is then set up, based on the amount and duration of data needed to be transported through the network in the form of ATM cells. The circuit is torn down when the session—the transmission—is complete.

The ATM switch provides dedicated circuits or paths dynamically or "on the fly" between users at each end of the network. ATM networks provide dynamic bandwidth, giving applications as much network resources as they need. ATM provides scalable speeds; the lower limit is determined by balancing the cost-effectiveness of ATM speed against the 5-byte header of each 53-byte ATM cell. In practice, this means speeds from 25 to 155 Mbits/s in the LAN, through 622 Mbits/s and up to 2.5 Gbits/s in trunk networks.

ATM Route Servers

In a native ATM LAN, the route server addresses level 3, or network layer, requirements in the ATM LAN environment. It allows network layer traffic to be routed from one level 3 domain to another when two ATM LANs are interconnected. For example, where two ATM LANs represent different network layer domains and are interconnected by ATM WAN links, protocol traffic from one ATM LAN to another passes through the route server which is responsible for establishing the virtual circuits between the ATM LANs. The route server also allows information to be routed between network layer addresses in the same ATM LAN, associating network layer packets with ATM addresses. A typical approach to the requirements of route serving has involved a router with an ATM interface connected directly to an ATM LAN switch.

The multicast server is responsible for implementing multicast with an ATM LAN. ATM LANs are connection-oriented, while conventional LANs are connectionless by nature. While it is easy to implement a multicast or broadcast in a connectionless LAN environment, this is not easily achieved in a connection-oriented environment. In a connection-oriented ATM environment, any transmission between two nodes requires a virtual circuit [whether a permanent or switched virtual circuit (PVC or SVC)] to be present between the two devices before traffic can be sent between them. The implementation of an ATM multicast therefore requires a one-to-many link or multiple links to be established.

An Overview of Switching

Switches arrived just as the market for bridges was dying—this was not a coincidence. Bridges were limited as they could support only two LAN segments at a time. Although switches vary (see section on cut-through vs. store-and-forward switching, to follow), many are really multiport bridges, making a whole LAN operate at the MAC layer, and therefore nonhierarchical. Switches are available for LAN types:

- Ethernet
- Fast Ethernet
- Gigabit Ethernet
- Token ring
- FDDI
- ATM

We mainly examine Ethernet switching here, as it is in Ethernet and Fast Ethernet LANs that the vast bulk of switches are being installed today.

Switch Types

There are two basic types of LAN switch: static and dynamic. The static switch enables the network manager to simplify tasks such as additions, moves, and changes to an organization's network by automating these operations in software. For example, a static switch would let the network manager move a user from one department to another. They would do this by calling up a representation of the hub on their monitor and using their mouse to assign the user to a new port by pointing and clicking. After this has been done, any data packets would be directed to their correct destination. Such changes take minimal time and physical resources.

The dynamic switch increases the bandwidth available to devices on the network by identifying the data packet's source and destination port addresses and opening a dedicated 10-Mbit/s circuit between them. In essence, a dynamic switch creates a private LAN by allocating bandwidth on demand. With traditional shared LANs, the packets would be forwarded to every device on the network. With switched LANs, the packet goes only to its intended destination. Technically this should improve security by preventing eavesdropping or sniffing. Many private LAN circuits could be open simultaneously, although once the packet has been sent, that particular circuit is simply dropped.

Both the dynamic and static switch categories can be subdivided into port and segment switching. In a dynamic port switch, each port connects to a single end station or server, making it most suitable for boosting bandwidth economically for users who require access to data-intensive applications. Standard network interface cards, driver software, cabling, and applications can be used without change. In other words, the prevailing LAN infrastructure is protected.

A dynamic segment switch works in the same way, but each port can connect to a whole network segment rather than just one end station or server. This allows the network manager to reduce a highly congested LAN to a series of smaller LANs, a function which was traditionally provided by bridges and routers.

A static port switch includes software which allows the network manager to move users from one shared LAN bus to another. Individual ports may be selected and assigned to different shared buses, helping to optimize port usage by reducing the number of modules that must be purchased for one hub. On the other hand, a static module switch allows the network manager to move an entire hub module, complete with all its ports, from one shared bus to another.

Cut-through versus Store-and-Forward

Switching basically operates at layer 2 (MAC layer) of the OSI reference model. There are two main methods of switching: cut-through and store-and-forward. With *cut-through*, the switch reads the header on the incoming data packet for a destination address and sends the packet straight on to the relevant port. One of the drawbacks of this method is that even defective or empty packets will still be forwarded. However, it does provide low latency. A cut-through switch forwards a frame to its destination, before it has itself received the complete frame, without error checking.

Some manufacturers still claim that cut-through is the only true method of switching. However, this type of switching can occur only between ports of equal speed unless congestion control tools are added, a major consideration for network managers who plan to implement high-speed networking in the foreseeable future. Also, simply installing a cut-through switch takes no account of any defective equipment which may be causing the network to be flooded with data error packets in the first place.

A *store-and-forward* switch reads the whole packet and checks its validity before sending it on to its destination. This has the advantage of removing non—bona fide packets but does incur slight delays in data transmission through increased latency. Store-and-forward switches are necessary to move frames from a low-speed LAN to a high-speed LAN.

The increased use of high-speed processors in most store-and-forward switches has helped to reduce latency times, while the added functionality offered by high-speed processors includes the capability to support virtual LANs and filter protocols, network statistics, and other superfluous data before forwarding the packet. In particular, claims about latency from manufacturers should be studied closely. There is no standard measurement for latency, and most proprietary published figures are likely to refer to packets being forwarded within the same module and without other traffic passing through the switch. Also, latency may be increased in chassis-based switches as a result of any delays incurred as the packet crosses the backplane.

How Switching Works

A switch divides a LAN into segments and a large volume of traffic into smaller traffic flows, cutting out competition for bandwidth. Many people believe that switches can replace shared hubs, bridges, and routers. Others see switches and routers as members of the same team, communicating over a high-speed ATM backplane.

Switching technology has to address several key issues. It must reduce traffic bottlenecks by stopping unnecessary traffic from crossing segments. It must afford multiple and simultaneous communication paths between segments. It must also meet manufacturers' claims that LAN switches allow parallel communications to occur at 10 Mbits/s without performance degradation. For example, a six-port switch should offer a matrix of routes between all of its ports, allowing three paths to be opened up simultaneously. Each path would operate at 10 Mbits/s, giving the network an aggregate capacity of 30 Mbits/s.

A switch learns on the job; that is, it examines all packets which come into its ports, reads the source MAC address, and logs that address in a table. Once it knows which addresses belong to which port, full switching functions can commence. If a packet comes in from an unknown address, the switch will broadcast that packet to all its ports, and when it receives a reply, it will know where the new node lies.

Switches are transparent to networking protocols, so they require minimal software configuration or, indeed, hardware upgrades. This means that network managers can use switches to limit the forwarding of, for example, IPX (Internet Packet Exchange) broadcasts to only IPX workstations.

Although a switch allows the creation of a series of individual collision domains (segments), the domains are all still members of the same broadcast domain and should still communicate with each other.

Although a basic switch actually provides a maximum bandwidth of only 10 Mbits/s between two nodes, that should still be enough to bring significant improvements in terms of network performance—ideal for small workgroups or for a directly connected node with its own private segment.

In a typical client/server application, however, large numbers of users will usually require access to the same server or set of servers. In that case, the segment on which the server resides could become a bottleneck in its own right. To obviate this eventuality, many switches now feature a high-speed uplink. Usually, this is a Fast Ethernet or FDDI port running at 100 Mbits/s, allowing the network manager to place the server on a 100-Mbit/s segment, creating a high-speed pipe which is fed into by the various dedicated 10-Mbit/s connections on the other side of the switch. This would remove the bottleneck with a 10-fold increase in bandwidth to the server. The advent of gigabit switching, manufacturers say, is because Fast Ethernet is now being installed to the desktop, and Gigabit Ethernet backbones will relieve the bandwidth bottlenecks that would otherwise occur in that situation, similarly.

Address Issues

Switches forward data packets between users according to an internal reference table of MAC addresses. Usually the size of the table for any one port or group of ports is commensurate with the amount of content-addressable memory. With switching, addresses are not aged as they would be, for example, in bridging. Instead, when the address table is full, the oldest address is simply removed to make room for a new one. Or, if an end station changes location, the first packet sent from the new location notifies the switch that it has been moved and its old address will be removed from the table. Even so, if a device is removed completely from the network, its address must still be manually deleted from the table. In instances which require a large amount of address learning and aging, the introduction of switching could help to boost performance slightly.

A segment switch can hold several addresses per port and can be used to boost performance between workgroups. In contrast, a workgroup switch will support only a small number of addresses per port and can be used to boost performance between specific devices on the LAN. Ideally, a workgroup switch can replace a hub shared by power users or in a situation where a server is accessed as a central resource. On the other hand, a segment switch is most appropriate as a link between workgroup switches and/or hubs.

An intelligent switching hub allows multiple LAN segments to communicate with each other. With the definition of a virtual workgroup, comprising separate devices on the same or different switched segments, the switching hub can filter the traffic between those end stations, boosting performance and potentially providing an extra layer of security. Such a hub could also serve as a high-performance collapsed backbone, connecting several segments from individual workgroup hubs. This offers benefits in terms of the centralization and simplified management of multiple LANs.

The careful integration of FDDI in a switched LAN is another way in which switching can boost performance. Generally, this is achieved via an FDDI module in the switch itself which offers a high-speed connection to servers and high-end workstations. By using the switching hub as a concentrator linking into an FDDI backbone, it is possible to create a switched network without having to invest in any additional LAN infrastructure.

Most manufacturers claim that their switches can operate on all ports at full wire speed. However, this depends on the allocation of processor power per port. Without sufficient power, it does not matter how many MAC addresses a switch claims to support. Even with sufficient power, bottlenecks are not precluded as data from a number of segments could be destined for one port. Some switches have a jamming process which prevents the arrival of any more data at the port in the event of a bottle-neck; however, in segment switches, this will also prevent communication anywhere else within the segment.

Layer 3 Switching

So far, only MAC-level switches have been dealt with here, and routing has been assumed to take the layer 3, or network layer, work. But in fact there are now hybrids—switches that incorporate limited routing. LAN switching will not take the place of routing. Routing remains at present the most cost-effective and technically efficient method of inter-LAN and LAN-to-WAN communication. Rather, we will continue to see the integration of switching and routing technologies in higher-level products. For a VLAN to operate throughout a network, it may be necessary to introduce such switches to break the network into manageable pieces without losing the advantages of the virtual network. Routers will still operate at the edge of the network to interface to the WAN—level 3 switches cannot offer the full security that a router can if properly configured.

VLAN Construction

Network upgrades and expansions are done in different ways depending on the company's culture and strategy, the cost, the difficulty, and the numbers of available staff. They are sometimes done gradually, by piecemeal replacement of physical segments, or like buses, they all come along at once, and a whole building is upgraded wholesale. Sometimes radical change is necessary; for example, a 10Base-2 (i.e., coaxial-cable) LAN with repeaters and bridges in a bus-based topology (see Chap. 3 for details) needs to be turned into a star-topology 10Base-T before doing any other kind of improvement if there are to be any new types of equipment installed ever again—while coax (coaxial cable) is easier to manage, it is not able to support new, faster networks or switches. While doing this major work, the company will obviously take the opportunity to install the extra bits and pieces to make it all work well, rather than simply just work. Thus, in this scenario, what is known as a "forklift" upgrade (whip out the old and shove in the new very quickly) is the best way, although it can be done one floor at a time or even one segment at a time. On the other hand, consider a network with good cabling and a LAN broken up into physical segments by routers: this can be upgraded gradually by adding a switch and VLANs and moving the users into them as is convenient, the way a new server is often added.

We have already established (in Chap. 1) what a virtual LAN is. In summary, we can say that it is basically a limited broadcast domain, meaning that all members of a VLAN receive every broadcast packet sent by members of the same VLAN but not packets sent by members of a different VLAN. We will move on to look at the problems of implementation later (Chap. 8). In this chapter, we will look in more detail at the various different types, all claiming to be VLANs.

The easiest way to understand VLAN is to compare it to a physical LAN—a collection of end stations, bounded by routers, which share a common physical connection: the cable. A VLAN is a collection of end stations which directly communicates without the need to traverse a router. Any device, anywhere on the network, can be a member of any virtual LAN, regardless of where other members of that VLAN are located, as long as the switches to which those end stations are connected support VLANs. Each end station on any given VLAN—and only those end stations—would hear broadcast traffic sent by other VLAN members and no other traffic.

VLANs are mechanisms used to determine which end stations should receive broadcast traffic, since it should not be sent arbitrarily to every connected user. This is important because broadcasts are a popular method for establishing communications between servers and groups of end stations by using such noisy protocols as ARP and IPX. In a bridged or switched network without level 3 segmentation, by means of either a router or VLANs, every end station listens out for every packet. If it is not intended for that station, it ignores it. From the network perspective, the whole point of a VLAN is to keep broadcast traffic from flooding the large flat network which is created by a series of layer 2 switches—really bridges. This means that in any medium-size network, it is common to have "broadcast storms"—or networks so flooded with traffic that they become congested and slow down or even crash on a regular basis. This, of course, was why routers were introduced into the centers of LANs in the first place—to create IP subnets and stop the broadcast storms. Now that routers are much slower than the speed required by today's LANs, and also with a need to give individual users dedicated bandwidth, the routers are giving way to switches, as we have already noted. Switches with VLANs can limit broadcasts by preventing them from propagating across the entire network.

VLANs should be seen as a solution to at least one of two problems:

■ Containment of broadcast traffic to minimize dependence on routers (as mentioned above)

- Reduction in the cost of network moves and changes

A smaller organization, in which broadcast traffic is not yet a problem or the cost of network moves and changes is manageable, may not want to bring in VLANs for the time being, especially given the present unsettled state. However, many larger networks are now experiencing one or both of these problems.

In organizations that are already replacing routers with switches (switches are cheaper and faster than routers, after all) and may soon face broadcast traffic containment issues, another element of the network architecture should be considered: the degree to which the network has evolved toward a single-user/port-switched LAN architecture. Although this is the stated aim in moving toward switching, it is more common at present for small groups of users to share a switched port—a hangover from the days of hubs and bridges, and why not, since a single user does not need dedicated bandwidth. . .yet. If the majority of users are still on shared LAN segments, the ability of VLANs to contain broadcasts is greatly reduced. If multiple users belonged to different VLANs on the same shared LAN segment, that segment would receive broadcasts from each VLAN, thus defeating the goal of broadcast containment. It is therefore imperative that a company plan its networking strategy carefully, and keep the possibility of implementing VLANs in mind during that process.

Once the need for VLANs has been determined, the planning process must also include the questions of server access, server location, and application utilization. It is important to determine the nature of traffic flow in the network. This should throw up the answers about where VLAN broadcast domains should be deployed, what role, if any, ATM needs to play, and where the routing function should be placed. In most cases, routers will be used mainly to keep the WAN traffic securely bounded, and therefore at the edge of the network, but it is also a good idea for routers to be able to handle the internal traffic, at least at first. This way, migration to VLANs can be performed gradually and the switches can learn addresses automatically (at least in some VLAN schemes).

The analysis of network traffic, applications usage, server access, and so on that is necessary in the VLAN migration process, and which will be greatly furthered by the implementation of RMON2 (see Chap. 6), may simply produce VLANs that correspond to functional teams or departments. At the moment, this does seem to be the way that idea is going, which is a pity, since it is a one-eyed view of the network. If migration is instead undertaken with a holistic view of the capabilities of VLAN technology—with the network designers asking "Who *should* talk to whom?" rather than "Who *is* talking to whom?"—it may become apparent that fundamental process and organizational changes are needed. In Chap. 2 we examined some of the changes that corporate structures are undergoing, with a need for increased flexibility in people leading to a similar need in the networking structure—flatter hierarchies, better process control, personnel constantly shifting—all these need a better-organized networking structure, and the implementation of VLANs is one of the means of achieving this.

VLAN Types

VLAN membership can be defined in several ways. VLAN solutions fall into seven general types, arranged in order from the simplest to the most complex:

- Port-based
- Protocol-based
- MAC-layer grouping
- Network-layer grouping
- IP multicast grouping
- Combination
- Policy-based

The different types can be of best advantage to users, network managers and administrators, or to a company as a whole [in the way it integrates its MIS (management information system) policies into its business]. There are also issues of manual versus automatic VLAN configuration, the bringing in of remote workers, and techniques by which VLANs may be extended across multiple switches in the local network and across the WAN.

Port-Based VLANs

Many early VLAN implementations defined VLAN membership by groups of switch ports (e.g., ports 1, 3, 5, and 7 on a switch make up

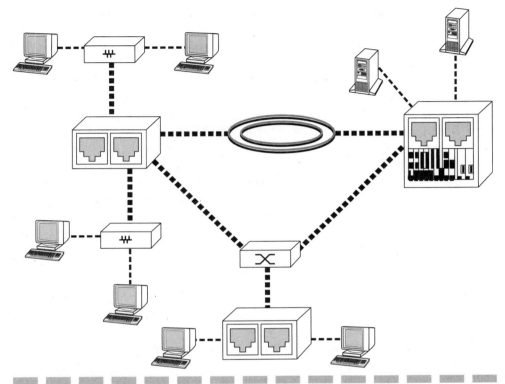

Figure 4-1 VLANs defined by port group (Source: Xylan Corp.)

VLAN A, while ports 2, 4, and 6 make up VLAN B). Also, in these early implementations, VLANs could be supported only on a single switch.

Second-generation implementations support VLANs that span multiple switches (e.g., ports 1 and 2 of switch X and ports 3, 4, 6, and 7 of switch Y make up VLAN A; while ports 3, 4, 5, 6, 7, and 8 of switch X combined with ports 1, 2, and 5 of switch Y make up VLAN B). This scenario is depicted in Fig. 4-1.

Port grouping is still the most common method of defining VLAN membership, with all vendors supporting it (usually in addition to more complex schemes) and configuration is fairly straightforward. With VLANs, there is often tradeoff between ease of configuration and ease of management. In this case, where the configuration is easy, management, although easy if the management software is good, is time-consuming. A damaging limitation of defining VLANs by port is that the network manager must reconfigure VLAN membership when a user moves from one port to another. In addition, defining VLANs purely by port group does not allow multiple VLANs to include the same physical segment (or switch port). Devices are assigned to virtual LANs based on the ports they are physically attached to; VLAN port assignments are static and can be changed only by the MIS staff. While they are easy to set up, they can support only one virtual LAN per port. Therefore, they have limited support for hubs and large servers, which generally need to have access to more than one virtual LAN.

These VLANs are definitely administrator-based, in that they are designed only to make the network run more efficiently (rather than to satisfy business requirements), particularly in the matter of flood prevention. Because of the manual changes needed, they do not, however, save the administrator a lot of time. An advantage is that they are secure, since only the network manager can make changes. These are the simplest form of virtual LAN but also provide the largest degree of control and security, due to the need for manual assignment changes.

Membership by MAC address

In *MAC address VLANs*, all address resolution is at layer 2, based on the MAC address. It is here that the first sophisticated VLAN management GUIs (graphical user interfaces) were first designed, making it possible for network managers to see easily just what, and who, was in which VLAN. Often color-coded, these GUIs make moves, adds, and changes easier.

Since MAC-layer addresses are hard-wired into the end user's NIC, VLANs based on MAC addresses (see Fig. 4-2) enable network managers to move a workstation to a different physical location on the network and have that workstation retain its VLAN membership—a process that can be automated. Even so, the time-consuming configuration process is all manual with this type of VLAN, and this is causing it to lose popularity in the face of more automated schemes. Some vendors have got around this problem by using tools that create VLANs based on the current state of the network. This gives a MAC address-based VLAN for each subnet existent at the time of setup. Stations with unrecognized MAC addresses (that the VLAN manager tool cannot place, that is) get put into a catchall unconfigured VLAN until manually attached to their rightful homes. In the pure MAC-address VLAN type, it is common for all end stations to have the same network (IP or IPX) address, with a router to interconnect the different VLANs. (This is related to protocol VLANs.)

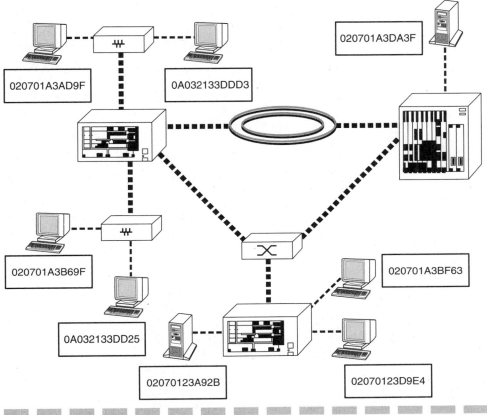

020701A3DA3F

020701A3AD9F

0A032133DDD3

020701A3B69F

0A032133DD25

02070123A92B

020701A3BF63

02070123D9E4

Figure 4-2 VLANs defined by MAC address (Source: Xylan Corp.)

This type of VLAN allows notebook and laptop computers to be plugged into the network anywhere and retain their memberships. In this way, a VLAN defined by MAC address can be thought of as a user-based VLAN. It is also not possible to have this feature on a layer 3 LAN, disadvantaging true remote workers, however (which many of the laptop users are), as the MAC addresses are not used across the network layer. It is in this type of VLAN that it is possible to have "overlapping" VLANs—that is, one user may belong to more than one VLAN at a time. In early MAC-address VLANs, although this was possible, it was not designed in as nobody considered the possible advantages, which are

■ Easy communications are possible between overarching business-need groups, such as sales directors and managers, who need to be both in sales and marketing VLANs.

■ It allows technical support staff to be on all VLAN membership lists.

This makes this type of VLAN a business-based scheme, giving business, not just network efficiency, benefits and is the precursor to the newer policy-based VLANs. Again, this overlapping facility may be useful, but it means a lot of time spent at the initialization of the VLAN system.

Protocol-Based VLANs

Another simple type of VLANs allows different groups to be set according to protocol (see Fig. 4-3)—for example, all users defined by MAC address forms one VLAN, others defined by their IP address forms

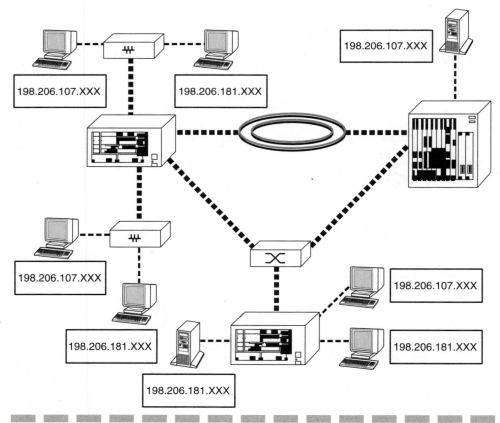

Figure 4-3   VLANs defined by protocol (Source: Xylan Corp.)

another, definition by IPX address another, and so on. A manager could decide, for instance, that a certain field indicating the protocol type within a frame will be used to determine VLAN membership, and then download this policy to all the switches in the network. For instance, this could be used to put all users who are using NetBIOS and IP into their own VLAN.

This is, however, a very crude way of doing it when it is used alone, but when used in conjunction with IP-based VLANs, it can allow otherwise nonparticipants to join in. It does have the advantage that many users could be in more than one VLAN at once, but this way of forcing VLANs onto a network has been abandoned by all serious vendors. It also has advantages for network managers running several protocols within a large network where distinct groups of users sit on different types already—here it can be profitably used to separate the different traffic types. Like port-based VLANs, these VLANs are administrator-based, in that they are designed only to make the network run more efficiently. Its greatest advantage is that it allows IPX networks to join the party—it can keep the SAP (Service Advertisement Protocol) broadcasts generated by IPX under control, in a simple way.

This is the simplest type of VLAN that operates at layer 3; on IP networks, this means that it can understand an IP address, and is therefore related to the IP-based VLANs examined next. Adding layer 3 intelligence to layer 2 switches allows the switches to look inside the packets to read the addressing fields. Precisely speaking, the switch performs a layer 3 lookup on an initial packet such as an ARP broadcast and maps the ID to the correct outgoing port, using the MAC address. The switch needs to be aware of all the protocols in use on the LAN, so care must be chosen in switch selection.

IP-Based VLANs

A more sophisticated layer 3 VLAN (see Fig. 4-4) is based on IP addresses or, less commonly, IPX network numbers. I have concentrated on IP as that is more common as Novell networks are migrated away from IPX.

IP-based VLANs are based on layer 3 information and use network (i.e., IP subnet) addresses. They also take into account protocol type (if multiple protocols are supported) in determining VLAN membership. Although these VLANs are based on layer 3 information in the packets sent, this does not constitute routing and should not be confused with a routing function. It can make use of the same routing tables, however. To use all

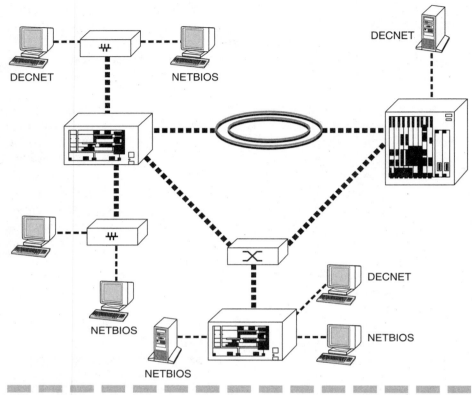

Figure 4-4   Layer 3 VLANs (Source: Xylan Corp.)

the facilities that such VLANs permit, there need to be layer 3 switches in the network that have this routing function. A router will also work to send packets between subnets while still retaining VLAN information, but this will create bottlenecks. Layer 3 switches usually have the packet-forwarding function of routing built into ASIC chip sets, which greatly improves performance over CPU-based routers. Migration of part of the network at a time is possible by having both layer 3 switches and routers operating at the same time in the network. Eventually, the router can be retired, when all network users are brought into the VLAN scheme.

Even though a switch inspects a packet's IP address to determine VLAN membership, no route calculation is undertaken, RIP or OSPF protocols are not employed (some layer 3 switches do not understand these protocols, anyway), and frames crossing the switch are usually bridged according to the spanning-tree algorithm. So from the point of view of a layer 3 switch with layer 3-based VLANs, the network with all its VLANs is still seen as a flat, bridged topology. The fact remains that no matter where it is located in a VLAN solution, routing of some kind is necessary to provide connectivity between distinct VLANs. The point must be made again about the difference between switches and routers. No matter how sophisticated the switch, it is not a router, which is a full-fledged "traffic cop" with intelligence as to addresses, and which acts as a firewall keeping the WAN at bay—most layer 3 switches do not have these capabilities yet. A fully switched secure network is possible, but separate security devices are then needed.

There are several advantages to defining VLANs at layer 3:

- It enables partitioning by protocol type. This may be an attractive option for network managers who are dedicated to a service- or application-based VLAN strategy, but see preceding text for proto-col-based VLANs.

- Users can physically move their workstations without having to reconfigure each workstation's network address—this is suitable only in pure TCP/IP networks.

- Defining VLANs at layer 3 can eliminate the need for frame tagging in order to communicate VLAN membership between switch-es, reducing transport overhead. IP-based VLANs, however, can suffer from performance problems (see Chap. 8).

- The network manager can decide which subnets are in which VLANs with simple drag-and-drop in the VLAN manager, and all the users are assigned along with their subnets.

- If defining by IP subnets, new users joining can be VLAN-aligned automatically by virtue of the subnet the switch finds them in.

VLANs defined at layer 3 are particularly effective in dealing with TCP/IP, but less effective with protocols such as IPX, DECnet, or AppleTalk, which do not involve manual configuration at the desktop. These protocols, while they have a large installed base, however, are not normally used in new installations, since Intel-architected PCs (rather than Macintoshes) and IP (Novell has more or less admitted defeat on IPX) have come to dominate the desktop. In addition, layer 3—defined VLANs have particular difficulty in dealing with unroutable protocols such as NetBIOS. End stations running unroutable protocols cannot be differentiated and thus cannot be defined as part of a network layer VLAN. A mixture of protocol-based and IP-based VLANs can be successful, but is not likely to become a permanent solution.

IP Multicast Groups Used as VLANs.   Multicasts are unlike broadcasts (which go everywhere on a network) and unicasts (which are point-to-point). Multicast traffic is point-to-multipoint or multipoint-to-multipoint and is rapidly becoming an expected service for data networks. Multicasts can be used for videoconferencing, stockmarket quotes, and news feeds to the desktop—users can expect more applications to develop as the benefits of multicasting continue to be realized.

Multicast VLANs (see Fig. 4-5) are created dynamically by listening to Internet Group Management Protocol (IGMP). As users open applications which use multicasts, they will dynamically join the multicast VLAN associated with that application, and when they close the application, they will disconnect from the multicast VLAN. Multicasts are gen-

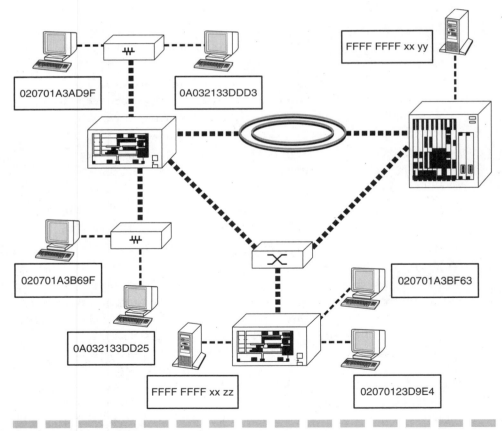

Figure 4-5   Multicasts used as VLANs (Source: Xylan Corp.)

erally steady-stream and bandwidth-intensive. Therefore it's best to put this type of traffic into its own VLAN to prevent network flooding.

There is a difference between what multicasts were designed for and what VLANs solve. Nonbroadcast multicasts were conceived as a way for MAC stations to ignore traffic which did not pertain to them, since some stations on the network cannot process packets at wire speed and there needed to be a way at the MAC layer that packets could be dropped before they consumed buffer space and CPU time. However, all multicast packets which are ignored by a station still use up bandwidth on the wire, and in a traditional bridged network, all multicasts are forwarded indiscriminately to all ports and networks. The goal of VLANs is to keep multicasts off LANs where none of the stations on that wire need them. This can be thought of as multicast packet limiting at a higher level in the hierarchy; while the point of multicast original was to limit multicast bandwidth into individual stations which did not need it, the point of VLANs is to limit multicast bandwidth out of switch ports into LANs which have no stations which need it. Even if that LAN consists of only one station, the more multicasts you can limit from being forwarded to that wire, the more bandwidth for unicast data you can have. (Since the point of switches is to have more internal bandwidth than each individual segment, dropping multicasts gives it room for sending unicasts which may be waiting for transmission to that segment.)

IP multicast groups are not really VLANs, as they are set up and broken down automatically for short lengths of time only. However, they do work insofar as they limit network flooding of broadcast packets, which is one of the main reasons for installing VLANs. In this way, they should be considered as a type of VLAN, even though if a company uses this, they may be able to put off having a "real" VLAN even if operating a totally switched network. This type of VLAN setup has nothing to do with the business of the enterprise, and all to do with the day-to-day performance of the network and the saving of the network manager's time. While this type of VLAN overcomes the scalability problem for a large enterprise, it is not useful for the logical organization of an enterprise's users.

Combination VLAN Types

A flexible definition of VLAN membership enables network managers to configure their VLANs to best suit their particular network environment. For example, by using a combination of methods, an organization that utilizes both IP and NetBIOS protocols could define IP VLANs corresponding to preexisting IP subnets and then define VLANs for Net-BIOS end stations by dividing them by groups of MAC layer addresses. This makes for a smooth migration to a VLAN setup and contains broadcasts from the beginning.

VLANs and ATM.   In a network with an ATM backbone but no ATM-connected end stations, ATM permanent virtual circuits (PVCs) can be set up to carry intra-VLAN traffic between multiple LAN switches. Two methods support such an implementation of the VLAN concept: MAC encapsulation and MAC translation. Both make use of the MAC layer definitions used in today's most popular LANs: Ethernet and token ring. It is important to remember that VLANs are based on MAC technologies which must be preserved even if the network includes ATM users. Once MAC packets are segmented into ATM cells, they cannot be interrupted until they are reassembled back into packets.

MAC encapsulation requires that all users of a single VLAN be of the same MAC type. Routers are required to interconnect VLANs of different MAC encapsulation types. The ATM Forum has defined two MAC encapsulation types: one based on the Ethernet framing standard; the other, on the token-ring framing standard. MAC encapsulation is best suited for networks which support only one MAC frame format: token ring or Ethernet. Where multiple MAC frame types are supported, MAC translation is probably a better answer.

With *MAC translation,* users of any MAC type can be in the same VLAN. Translation-based VLANs are preferred since they support any-to-any connectivity. Implementation of translation is more complex, so costs are higher than encapsulation-based products.

Usually, organizations that implement ATM backbones would also like to connect servers directly to those backbones. A server is, however, an end station, and as soon as any end station is connected via ATM, LANE must be introduced into the network to enable ATM-connected end stations and non-ATM-connected end stations to communicate, which adds a new level of complexity. Since routing remains necessary in any mixed ATM/shared media environment to forward inter-VLAN traffic, network designers are faced with the question of where to locate the router functionality. (See Fig. 4-6.)

In *edge routing,* traffic within VLANs can be switched across the ATM backbone with minimal delay, while inter-VLAN packets are processed by the routing function built into the switch. In this way, an inter-

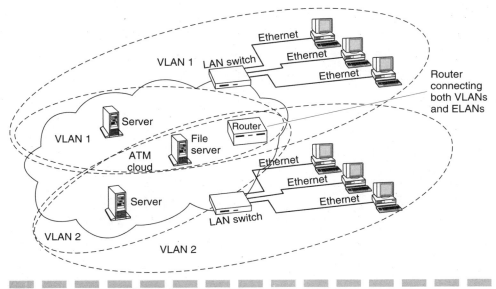

Figure 4-6 Router connecting VLANs and ELANs

VLAN packet does not have to make a special trip to an external router, eliminating a time-consuming extra hop. Edge routing will function in multivendor environments if each vendor's equipment supports LAN Emulation. The primary disadvantage of edge routing is the difficulty of managing multiple physical devices relative to having centralized management of a consolidated router or routing function. Additionally, edge routing solutions may be more expensive than centralized routing solutions made up of a centralized router and multiple, less-expensive edge switches.

The *route server* model breaks up the routing function into distributed parts. The same packet waits in the cache of the LAN switch at the edge of the ATM backbone before transmission. In this process, the packet itself never traverses a router. The only traffic to and from the route server is the signaling required to set up a connection between LAN switches across the ATM backbone. The advantage is that less routed traffic must be diverted to the route server, often reducing the number of hops required through the backbone. Also, overall traffic across the route server's one arm is reduced. (See Fig. 4-7.)

LANE. Deploying an ATM backbone also enables the communication of VLAN information between switches, but it introduces a new set of

Figure 4-7
Route server in a
mixed ELAN/VLAN
environment (Source:
3Com)

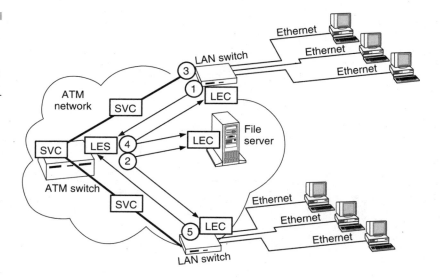

issues with regard to LANE (see Chap. 5). LANE in an ATM network produces Emulated LANs, or ELANS, rather than VLANs as discussed here, but with port group-defined VLANs, the LANE standard provides for a nonproprietary method of communicating VLAN membership across a backbone.

LANE under ATM automatically gives a facility for separate ELANs. These are not true VLANs as they do not necessarily have all the functionality that a good VLAN will have, such as a choice of type or a graphical manager. With LANE, a choice is possible between the type of LAN to be emulated—either Ethernet or token ring. These do not, however, use CSMA/CD or token passing. It is perfectly possible to create multiple ELANs and mix these types. The ATM adapters have LANE clients to become part of logical ELANs. It is also possible, with some vendors, to trunk through between the ELANs and other VLANs running on non-ATM switches. With LANE, no tag interchange mechanism is needed as each ELAN has its own "corridor." ELAN membership is learned from a storage table in management software.

With the introduction of ATM-connected end stations, the network becomes a truly "mixed" environment, with two types of networks operating under fundamentally different technologies: connectionless LANs (Ethernet, token ring, FDDI, etc.) and connection-oriented ATM. This environment puts the responsibility on the ATM side of the network to "emulate" the characteristics of broadcast LANs and provide MAC-to-ATM address resolution. However, from an administrative point of view, many organizations may not want to employ separate management software for the ATM backbone and may prefer to source both edge devices (LAN switches) and backbone devices (ATM switches) from the same vendor.

An ATM backbone can enable all end stations from multiple VLANs to access a centralized server or servers without passing through a router by establishing a separate ELAN for each VLAN (though some vendors support multiple VLANs with one ELAN). Since most traffic in a network is between client and server, establishing VLANs that overlap at ATM-attached servers greatly reduces the number of packets that must be routed between VLANs. Of course, there is still likely to be a small amount of inter-VLAN traffic remaining. Therefore, a router is still required for traffic to pass from one VLAN to another (and, therefore, from one ELAN to another).

Layer 2/3 VLANs. Virtual networking using layer 2/3 VLANs blends some of the qualities of traditional routing and connection-oriented technologies, sometimes with policy-based automated network management services. When deployed, virtual networking provides greater reliability and security than do routers and at a much higher level of performance. A virtual network appears to the user end station as a traditional layer 3 routed network, internally it appears to be a single LAN to any traditional router connected to it.

It is important to recognize the fact that a virtual network or VLAN is an isolated grouping of users. To communicate, those users must be in the same subnet or network address range. To communicate between VLANs, they need further assistance to communicate between VLANs. It is not possible to switch between VLANs, since this would collapse all VLANs into a single broadcast domain, i.e., a single VLAN. Routers or routing capability is thus needed to retain the separate broadcast domains by acting as broadcast firewalls, while at layer 3, allowing users on different subnets and networks to communicate.

Virtual networks can be used with or without traditional routers. Early deployments of virtual networks still include routers for WAN connections, while relying on switching for the LAN environment. Virtual networks appear as multiple logical subnets to traditional routers. From the router's viewpoint, the virtual network functions like a LAN at the physical layer. This allows routers to interoperate with virtual networks with no functional changes. This protects the existing investment in WAN infrastructure. Alternatively, a virtual network can also be used to interconnect a group of traditional routers. In this case, the virtual network appears to the routers as a single LAN, transparent at the net-

TABLE 4-1

Comparison
between Emulated
LANs, Layer 2
VLANs, and Layer
2/3 VLANs

VLAN type	Definition	Functionality
LAN	A single physical LAN segment	Same function as a hub or basic bridge, "true plug and play" out of the box; by itself, does not provide a lot of functionality or value to the network
Layer 2 VLAN	Many physical or LAN segments or users bridged together	Limits flooding, but has high emulated administrative workload, similar to going back from routing back to specialized filter bridges; static network configuration means user service demands will not be met; layer 2 VLANs bring the worst aspects of bridging to ATM users; there is no interconnectivity of layer 2 VLANs without layer 3 routers
Layer 2/3 VLAN	Virtual networking between LAN segments or users	Allows virtual networking without any layer 2 or layer 3 restrictions; preserves the reliability of layer 3 routing while leveraging the network benefit of layer 2 connection-oriented switching; centralized administration of access, bandwidth allocation, and connections; brings the benefits of ATM to LAN users without the need to upgrade hardware or software

work layer. Thus, the virtual network acts as a very high-speed switched LAN, providing a reliable interconnect between routers. Table 4-1 summarizes the different types of VLAN discussed.

Policy-Based VLANs

Policies have been used in network management for over two decades, but are still not widely implemented except in large companies (very small ones probably have little need). Where the access rights of multiple users to resources are of concern, a general policy framework is needed to control the behavior of the various agents within the network: users, administrators, software applications, and hardware infrastructure. The idea of a policy is well understood—they are house rules with possible actions, in effect. They work on an if-then structure. Policies can prohibit, permit, or obligate actions for both people and the hardware and software in the network. An example of the latter is a policy for the escalation of network events to network alarms. Where people and machines come together is, for example, where a policy sets the threshold for an alarm and another sets the policy for the resulting action by a network manager. Policies in this way cut across all management tasks. (See Fig. 4-8.)

Policies do not exist in a vacuum; they are contained in domains. In the example given in Fig. 4-8, of a threshold for alarms, the domain is the network, or a part of the network. If there are different criticalities in different parts, then thresholds can differ. Of course, domains can overlap—part of the physical network structure may be covered by one policy (e.g., any breakdown at all requires an engineer to attend within 30 mins), but the company may have a policy stating that all nonfunctioning devices must be fixed within 12 h. Obviously, in this case, there is no conflict—here it is a matter of going from the general to the more specific (usually good practice in anything). For automated software, this is not so obvious—a type of expert system is needed to sort out the priori-

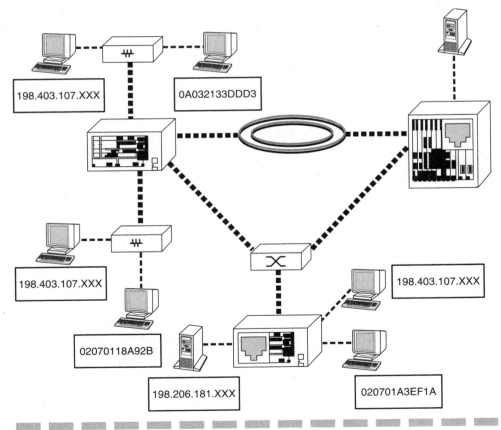

Figure 4-8 Policy-based VLANs (Source: Xylan Corp.)

ties. In other cases, especially where people and devices are brought within orthogonal policies, there can be conflict. For example, if a policy stipulates that the mail server be brought down for maintenance on Wednesday mornings at 10 A.M. for 2 h (as does happen with an ISP in the United Kingdom), but another requires that customer queries be answered by email within a certain time that is almost up by the time 10 A.M. comes around—what then? The software should include conflict-resolution strategies, but if all fails, it has to call for help—and that means more work for the hard-pressed network manager, who may not even be the best person to make a decision when it comes to policies. (The domains in question in the example are the network attached to the mail server and customer interaction.) Such conflicts can lead to unpredictable control actions and endless loops.

To get the management right, one needs to ask the correct questions at setup:

- What are the domains?
- What are the rules?
- What objects are there?
- What are their attributes?
- How should objects be grouped?
- What events trigger the policy driver?
- What actions result?

Hopefully, such sorting out at the start will help to avoid anomalous situations arising.

Policy management has so far been applied to certain classes of management software: fault, performance, security, configuration, and accounting. But it is beginning to be applied to other classes.

This is the most powerful VLAN implementation. It lets network managers use any combination of VLANs policies to create VLANs that suit their needs. They assign devices to virtual LANs using policies listed above which will be applied to all switches. Once a policy has been downloaded to a switch, it is applied evenly throughout the network and devices are placed into VLANs. Frames from devices are constantly reevaluated so that VLAN membership can change according to the type of traffic that a device is generating.

Policy-based virtual LANs can implement a variety of assignment methods, including any of the VLAN types listed above: MAC source addresses, layer 3 addresses, and protocol type field. It's also possible to combine different policies to make a policy that specifically meets a network manager's needs.

A few VLAN vendors emphasize the policy-based nature of their products. This shows in such cases where security is set by policies, perhaps ensuring that any communication between nonsecure and secure VLANs must pass through a router. The VLANs constructed by policies are still broadcast domains (a particular class of the more general domains discussed above), but membership of the domain in a policy-based VLAN is more granular, based on any criterion or criteria within a packet, and with multiple VLANs per switch port.

Potentially, policy-based or filter-centric VLAN membership gives great flexibility, but if implementors are not careful, it can be the cause of the introduction of greater complexity and administrative overhead. Technically, policy-based VLANs are the result of applying frame-filters to broadcast domains, which allow such sophisticated VLANs to be created. While policy-based VLANs allow more granular control of traffic, it means that the LAN switches in the broadcast domains must support proportionally more VLANs per switch than do simpler layer 2 switches.

A further VLAN innovation uses a superset of the filter-centric method for constructing VLANs. Rather than configuring the various filters needed to create VLANs by hand, subnet-centric VLANs automate the process, and automatically map the relationship between devices and their layer 3 network or host address to a VLAN. This technique is sometimes called "creating a relational LAN." The main advantage of this approach is there is no manual VLAN administration involved for certain protocols.

Policy-based VLANs are more dependent on specific network management than are other VLANs (see Chap. 6 for more detail).

Automation

Throughout I have emphasized the balance between automation (easing the network managers' workload), and flexibility and functionality (which may further the business aims). Central to VLAN deployment, therefore, is the amount of automation in the configuration. As we have seen, to a certain extent, the degree of automation is correlated to the type of VLAN; but the specific solution will play its part. There are three primary levels of automation in VLAN configuration, as in Table 4-2.

Manual configuration enables a high degree of control but is seldom practical in larger networks, and defeats a primary purpose of VLANs—

TABLE 4-2

VLAN Automation

Type	Configuration	Changes
Manual	Initial setup is manual	All changes are manual
Semiautomatic type 1	Initial setup is manual	Subsequent changes are automated
Semiautomatic type 2	Configuration is automatic	All changes are manual
Automatic	Configuration is automatic	Subsequent changes are automated

easing moves, adds, and changes. The ease will depend on the vendor's interface. The type I have called "semiautomatic type 2" shares that limitation, but tools are there for the initial setup. This type can have problems with later changes. Semiautomatic type 1 will take time to start but is thereafter easier to handle, and there will be no problems with later changes. This is a type suited to a gradual migration. A fully automatic VLAN will be the easiest but the least flexible, in that it gives no choices—a network manager's dream, it will at least make for an efficiently run network; but if using DHCP, there could be conflicts—see the next section.

Communicating VLAN Membership Information

Switches must have a way of understanding VLAN membership—in other words, which stations belong to which VLAN—when network traffic arrives from other switches, or VLANs would be limited to a single switch. In general, layer 2—based VLANs (i.e., defined by port or by MAC address) communicate their VLAN memberships explicitly, while for IP-based VLANs with IP subnets, it is implicitly communicated by the IP address. Depending on the particular vendor's solution, communication of VLAN membership may also be implicit in the case of layer 3—based VLANs in a multiprotocol environment. These different methods are of course implicit in whatever choice of VLAN is made, and results automatically in a user-based or a network-manager-based VLAN. The choice is not really up to the customer; at the present stage of development, it is best to buy from whomever has sold you the switches, given the lack of interoperability at the moment. If the switches in the network have not yet been bought, it is best to choose according to other criteria than what VLAN type the vendor sells—VLANs are developing all the time at the moment. Therefore, if deciding to opt for

VLANs, the customer has little choice about the type, except within the range offered by a vendor chosen for non-VLAN reasons.

Interswitch Communication

Three methods are used for interswitch communication of VLAN information across a backbone (not including ATM backbones): frame tagging, signaling, and time-division multiplexing. LAN Emulation under ATM also gives a sort of VLAN configuration. Tagging is the method that all current protostandards are concentrating on, and thus the method that vendors are busy designing at the moment.

VLAN Trunk Tagging. Trunking is the mechanism by which the devices within the network read and understand the virtual geography of the network—when they have this information, trunking can take place. This allows the outgoing interface card to know to which local ports to flood a broadcast. This means that any VLAN, once tagged, will be able to receive traffic intended for it no matter where its origin without the need for a router. This will also extend VLANs across the WAN to remote users and between distant offices.

Since each frame is received from an attached device, the switch has to use some type of VLAN membership policy to determine which VLAN ID tag to attach. It is by means of this tagging that the switches know which VLANs lie where, not which users are in which VLAN. IEEE 802.1Q is an emerging trunking standard due out later in 1997 (at time of writing). It enables switches, routers, and servers from multiple vendors to interoperate across high-speed links with knowledge of VLAN membership (see Chap. 5 for details). IEEE 802.1Q will provide a mechanism for identifying individual virtual LANs over different LAN media.

Trunk tagging differs from internal tagging, which aids determination of VLAN membership within a single switch. Trunk tagging obviates the need for switches and routers to reanalyze header information. Currently, 802.1Q does not address VLAN identifiers for VLANs running over FDDI, Fast Ethernet, or Gigabit Ethernet. Meanwhile, Cisco has extended the existing IEEE 802.10 metropolitan area network (MAN) security standard to allow it to act as a VLAN trunking mechanism over FDDI and intends to extend 802.1Q to Fast Ethernet and Gigabit Ethernet. Cisco's VLAN trunking protocol for Fast Ethernet is called *Inter Switch Link* (ISL).

In the case of broadcasts and traffic with unknown destinations, the switch needs to know which subset of ports to flood the frame out on.

For unicasts, the switch needs to know which bridge forwarding table to consult to forward the unicast frames to their intended destination. Since each VLAN has a different set of members and some members are simultaneously resident in multiple VLANs, a separate bridge forwarding table is required for each VLAN. However, if a single forwarding database is used, one per network is all that is necessary.

Signaling. Layer 2 switches, like bridges, have cached address tables. Like bridges also, they are not intelligent devices, which means that the tables tell it whether the traffic is intended for a port of its own. If the traffic is destined to go outside its own domain, it will go to a router, or a layer 3 switch with routing tables. These understand not only its own area, but also other routers in the network or outside it, based on IP addresses. When an end station broadcasts its first frame, the layer 2 switch uses the tables to resolve MAC addresses. Where there are many switches, the constant updating information that they pass among themselves causes so much backbone congestion as to negate the benefits of VLANs, so this method is not scalable and not practicable.

Time-Division Multiplexing. The third method for interswitch communication is by *time-division multiplexing* (TDM), which works the same way on the interswitch backbone to support VLANs as it does in the WAN environment to support multiple traffic types except that here, channels are reserved for each VLAN. This way, the traffic is slotted together along the same pipe with each stream remaining separate. Although this approach does cut out some of the problems inherent in signaling and frame tagging (see Chap. 8), it also wastes bandwidth, because a slot dedicated to one VLAN cannot be used by another VLAN, even if that channel is not carrying traffic at that time. On the other hand, some people are saying that in the future we will not have to worry about bandwidth; it will be so cheap that the answer to any problem will be "throw bandwidth at it." But we still want efficient networks, and it seems that all available bandwidth will be used as new and bigger applications arise to make use of it.

VLAN Issues

VLANs and DHCP

With the Dynamic Host Configuration Protocol (DHCP), there is a different way to reduce the workload associated with administration of IP addresses. Unfortunately, DHCP can actually conflict with VLAN implementation, especially with layer 3, IP-based VLANs.

DHCP dynamically allocates IP addresses to end stations for fixed periods of time (see Chap. 5 for details). If an IP address is no longer appropriate for any end station, the server allocates it a new IP address. In this way DHCP can enable workstations to be moved from subnet to subnet without the network administrator or the switches having to manually configure IP addresses or update table information.

A network manager can, with DHCP, implement logical workgroups by range of IP addresses, in a similar way to IP-based VLANs. However, these workgroups are not the same as an IP-based VLAN in that the members of the groups are still bound by their physical subnet to their location, although there can be multiple groups residing in each subnet. DHCP implementation may reduce the administration overhead of TCP/IP networks in this way, but DHCP cannot control network broadcasts as VLANs do, which is the main reason at present for implementing them. DHCP is also applicable only to IP, and therefore any non-IP clients (such as IPX, or DECnet) would be left outside any grouping. However, for smaller, pure-IP networks, DHCP may be enough.

Conflicts can arise with addressing—when users move with their own computers to a new subnet, the DHCP server will dynamically allocate a new IP address for that workstation. Although this is for a limited time, the new address given will always be in the new range. Yet, this workstation's VLAN membership is based on the old IP address in the old range. The network administrator would have to manually update the client's IP address in the VLAN tables in the switch by means of the VLAN manager tool. This would eliminate both the primary benefit of DHCP and one of the primary benefits of having a VLANs. While this seems to mean it is impossible to have both VLANs and DHCP, some vendors are working to incorporate DHCP into their VLAN schemes. Presumably, this will mean that any DHCP updating will also automatically change the VLAN membership tables.

While MAC-address-based implementation can be made to work with DHCP, the benefits of each are lost similarly. There is much less chance of vendors being able to tie the two together than in the case of IP-based VLANs. Port-based VLANs can be used together with DHCP, and these can be complementary, but port-based VLANs are, as noted earlier, only the first step into VLANs and are not likely to persist in the face of the more sophisticated schemes that are becoming available.

Because of the trends toward server centralization, enterprisewide email, and collaborative applications, various network resources will need to be made available to users regardless of their VLAN membership. Ideally, this access should be provided without most user traffic having to traverse a router. In nonoverlapping VLANs, this is not possible—a router or a layer 3 switch with routing functionality must intercede, causing bottlenecks. Given the increasing preponderance of IP-based VLANs, because of their automatic status, fewer and fewer VLANs are able to operate on an overlapping basis. This has implications for use of shared resources such as mail, print, and fax servers, and printers (depending on the organizational structure of such devices within the network).

Organizations that implement VLANs recognize the need for certain logical end stations (e.g., centralized servers) to communicate with multiple VLANs on a regular basis, either through overlapping VLANs (in which network-attached end stations simultaneously belong to more than one VLAN) or via integrated routing that can process inter-VLAN packets at wire speed. From a strategic standpoint, these organizations have two ways to deploy VLANs: an "infrastructural" VLAN implementation or a "service-based" VLAN implementation. The choice of approach will have a substantial impact on the overall network architecture, and may even affect the management structure and business model of the organization.

Functional Grouping

An infrastructural approach to VLANs is based on the functional groups (the departments, workgroups, sections, etc.) that make up the organization. Each functional group, such as accounting, sales, and engineering, is assigned to its own uniquely defined VLAN. According to the 80/20 rule, the majority of network traffic is assumed to be within these functional groups, and thus within each VLAN. In this model, VLAN overlap, when possible, occurs at network resources that must be shared by multiple workgroups. However, it is widely thought that the 80/20 rule has now been irretrievably broken. With the advent of Internet access for many end users (depending on company policy and user status) and increasing intranet access, access to WAN traffic is becoming far more frequent than before. Especially in an intranet, the users have no knowledge of where the information, file, or other resource is located and if they did, would not care. The point of such a structure is, after all, to remove the idea of physical location.

Network Resources

A service-based approach to VLAN implementation looks, not at organizational or functional groups, but at individual user access to servers and applications—that is, network resources. In this model, each VLAN corresponds to a server or service on the network. Servers do not belong to multiple VLANs—groups of users do. In a typical organization, all users would belong to the email server's VLAN, while only a specified group such as the accounting department plus top-level executives would be members of the accounting database server's VLAN.

By nature, the service-based approach creates a much more complex set of VLAN membership relationships to be managed. Given the level of most VLAN visualization tools presently available, a large number of overlapping VLANs using the service-based approach could generate incomprehensible multilevel network diagrams at a management console. Therefore, to be practical, service-based VLAN solutions must include a high level of automatic configuration features. In response to the types of applications organizations want to deploy in the future, as well as the shift away from traditional, more rigid organizational structures, companies may wish to implement such VLANs. However, many of the more automated VLANs at present do not allow overlapping VLANs, being IP-address based with DHCP (see Chap. 5) or other, proprietary, automatic allocation.

Network Operating Systems

Confusion between the function of a VLAN and the NOS (network operating system) arises because in some cases, it appears that they do the same job, especially where Directory Services help give services to users. The most common NOS is Novell's NetWare (now renamed to IntranetWare, but the installed base is still largely NetWare), which is the proud possessor of NDS (Novell Directory Services), which allows a single login to give access to all authorized network services regardless of server and geographic location. This came along with NetWare 4.0, but the version in 4.1 is the important one—it allows the administrator to merge directory trees rather than having to create a new tree every time and includes the ability to move and rename containers from the management console, which means that network managers can also reconfigure the hierarchies of directories without having to completely re-create the directory concerned. NDS is therefore a distributed database that gives all network users and resources status as objects on the network that can be viewed and managed from a single central location. This gives, of course, user access in the same way that VLANs can, thus the confusion. It does not, however, do anything to prevent broadcast storms. While users may not be able to see the data flowing past their computers, it is still there and causing congestion. Current VLANs, plus Directories, however, do give the much-needed flexibility.

There are other, similar services in other NOSs. Banyan's VINES (Virtual Networking System) was famous for its StreetTalk, for example, and this still has a large installed base, especially where integration with mainframes is important because of VINES' good support in that area, but VINES has waned in popularity in recent years, and few new systems are being installed.

Microsoft NT is the up-and-coming NOS, in many instances replacing NetWare. It does have Directory Services, but this feature is very restricted—there can be only one location per tree, meaning that not only must a tree be built for every location, but it is a nightmare trying to keep the multiple directories thus produced up-to-date. Where scalability is not a problem, this is proving popular. Novell's NDS supports NT, and a hybrid solution is sometimes adopted. NDS was also licensed for use by developers in 1996, and there may well emerge new integrations for this product. The latest enhancement to NDS is support for the *Lightweight Directory Access Protocol* (LDAP)—an emerging directory access standard for the Internet. This will give users the potential ability to find and use the resources in their Novell directory over Internet connections.

Equipment Used

The types of hardware and software to be found in a VLAN are the same as on any network:

- PCs, workstations, and servers equipped with NICs
- Cabling
- Switches at layer 2 and possibly layer 3
- Routers
- Route servers on ATM networks

- Hubs, where end users can be grouped—these will be replaced by switches
- Operating systems and network operating systems (these will probably, but not necessarily, be separate items)
- Network management and monitoring system(s)
- Protocol stacks

The difference is in the way the devices are deployed. A VLAN is basically a switched environment; routers will be used mainly at the edge of the whole network or internetwork. Hubs will be replaced very quickly by LAN switches, ports of which are getting cheaper by the day. Such hubs as are still used can only be at ends of the network where whole parts of the VLAN system can be grouped together as a non-VLAN within the greater network. Where switches are put in place, they must have software-configurable backplanes to deliver high speeds across multiple technologies.

In general, no replacement of NICs is needed, and routers can be redeployed as edge devices as migration to switches occurs. Cabling has to be category 5 UTP to get the best out of a switched network; that allows any technology at all. Assuming that a switched network, or at least a category 5 star topology network, is already in place, then the cabling does not need to change. The only new equipment needed is switches, plus the software to run the VLANs and manage them. This may entail replacing the network management software with the latest version.

The network management software must have a good GUI, where it is easy to view, and make changes to, VLAN memberships by dragging and dropping. It must also have a good view from the network point of view, also drag-and-drop action, and these two views must be integrated together so that changes from one affect the other. Some vendors' VLAN management tools work only on the switches that support VLANs, but it is a good idea, where gradual migration is to be employed, to use a tool that covers all switches on the network. It is also preferable to have the management tool integrated into the wider management system from the chosen vendor, and also into an "umbrella" system, which in most cases will come from a third-party vendor.

Network monitoring software using RMON (see Chap. 6) is absolutely necessary and often comes built into switches, but not all nine groups will usually be supported on all ports of the switches. Four groups is enough to gain a picture of the network, though seven provide better management tools.

Managing Switched Networks and VLANs

Switches are already being widely deployed to displace distributed back-bones, collapsed backbone routers, departmental routers for data closet microsegmentation, and shared-media hubs for desktop connectivity. While there are many compelling price/performance benefits of a switching environment, the transition to switched internetworking involves three fundamental structural transitions that have profound implications for the management of the network. As switches are used to provide more bandwidth through microsegmentation, a subnet that once mapped to a single hub can now be replaced by a more complex hierarchy of switches and hubs. End-to-end network management involves more physical devices, more types of devices, and more links.

Switches are often used to displace routers for microsegmentation. While switches are easier to administer and can significantly reduce network latency, they cause broadcast domains and subnets to expand and cover a much larger number of devices. Larger broadcast domains can reduce bandwidth efficiency by propagating broadcast traffic, threatening the overall stability of the network.

VLANs Decouple Logical and Physical Network

Partitioning of a flat switched network into routed subnets can be reestablished by creating VLANs and inter-VLAN routing. With configuration switching (port switching), for example, VLANs (i.e., virtual subnets) can even be defined down to the shared-media hub port level. Consequently, clients on any hub port can be assigned to either one or two VLANs, and a central server can now be part of, say, both VLAN A and VLAN B, completely breaking the previous linkage between the physical and logical network.

These structural differences pose some significant problems and challenges for managing the network. Management tools that were initially designed by umbrella network management system platform vendors or network equipment vendors for shared-media router/hub networks have serious shortcomings in a switched internetwork. For example, a tool that provides an Internet logical view of the network reveals only the routers, router interconnections, and subnets.

Where network maps continue to be constrained to the logical view, the portions of the network infrastructure which are invisible to the network administrator will continue to grow as more switches are deployed. If this radical divergence of the physical network from its visual representation in network management systems continues unchecked, the network administrator's ability to troubleshoot end-to-end connectivity or assess the potential impact of link failures in the switched internetwork will be seriously impaired.

In addition to preventing the flattening of switched networks, VLANs (as already discussed) have the potential to streamline the implementation of adds, moves, and changes of desktop connectivity. However, the daunting task of VLAN configuration and administration has caused most managers of switched internetworks to defer virtualization until better management tools are available. Because VLANs are essentially a logical overlay of the physical network, a high-productivity global VLAN configuration application must have access to a complete physical layer map of the network.

From these examples, meeting the challenges of managing shared and switched networks requires a new generation of applications specifically modified or designed for managing switched internetworks. A key enabler of many of these new applications is a flexible mapping mechanism which can integrate and unify the physical, logical, and virtual views of the network, regardless of the technologies used in its construction.

Multilayer Topology

Sometimes the terms *discovery* and *topology* are used interchangeably. This is unfortunate because it tends to mask the differences which may exist between the capabilities of different management applications.

Discovery is the best term to describe the process of making visible the various physical and/or logical entities of the network. Topology, on the other hand, is an appropriate term for the tabulation and graphical representation of the connectivity relationships that exist between these entities. Developing a topological map of the network, therefore, involves both discovering the network entities and retrieving information from MIBs (management information bases) within these entities which characterize device interrelationships.

As an example, the IP mapping functionality incorporated in most NMS (network management system) platforms (such as Hewlett-Packard's OpenView Network Node Manager and IBM's NetView for AIX) is limited to automated discovery of the routers and subnets within the network, plus automatic generation of a layer 3 topology (e.g., IP or IPX Internet) showing the router-to-router interconnections and router port-to-subnet relationships. In addition, some NMS platforms can autodiscover a simple listing of all devices (end systems and hubs) which participate in each subnet.

The key to managing switched internetworks will be a multilayer topology system which goes beyond discovery of the simple layer 3 topology by discovering all the entities in the network and fully characterizing their connectivity relationships via a physical layer topology and a data-link layer topology, as well as a layer 3 topology.

VLAN topology can be defined as an overlay of the physical topology. In addition to capturing relational information at each layer, the multilayer topology system must be capable of correlating the topology model at each layer with the models at the other layers. This allows the network administrator to select a region of the network and transform between topology views to examine it from each of the available perspectives.

Requirements for a Multilayer Topology System. Building a layer 3 topology is relatively easy because routers must be aware of each other in order to perform their basic function. Therefore, standard routing information interchange protocols and MIBs are adequate to capture and represent Internet topology.

A multilayer topology, however, is much more challenging to implement because now all the other internetworking devices (hubs, switches, bridges) must also be aware of how they are connected to the rest of the network. This enhanced awareness of connectivity requires that a high degree of intelligence and efficient algorithms be incorporated in embedded management agents, as well as in the management application system.

Another basic requirement is a mechanism to allow the agents in each device to exchange topology information with other agents in neighboring devices using efficient protocols which will not burden the network with management broadcast traffic. Each management agent must then maintain a table of local topology information in an MIB which can be retrieved via SNMP by the system building the global topology. The embedded agents must also be aware of VLANs and include this information in the local topology MIBs.

Topology/VLAN MIB. Intelligent agents store topology information in a common topology MIB format. The typical multilayer topology MIB would include

- Chassis and backplane type
- Link layer addresses for each port
- Layer 3 address(es)
- Slot and port numbers mapped to slot and port numbers of adjacent devices
- VLAN port groupings
- VLAN tags on physical links
- Timestamp indicating last topology change

In August 1977, a draft VLAN MIB based on SNMP2 was issued for use with the new VLAN standard. However, it deals mainly with port identification rather than the full list above.

Topology Database.   The global multilayer topology database must have a well-structured database schema to allow tight integration of logical, physical, and virtual views of the network. This requires sophisticated algorithms which are highly optimized for synthesizing information from multiple devices and technologies into a single relational model. In addition, a near-real-time updating mechanism must be provided to allow the database to accurately reflect topology changes when new devices are installed or reconfigured or an existing network element fails.

It can be seen that a multilayer topology engine is essentially a real-time, distributed processing system requiring close cooperation among many elements of the network. The full potential of this concept will be realized for multivendor switched internetworks when the standards are defined to allow interoperability among topology agents and topology-based applications. In the interim, the vendors of network infrastructure products must provide extended topology capabilities within their own proprietary VLAN and VLAN management implementations.

Multilayer topology information is the key enabler for integrating a wide range of switched internetwork management capabilities within an easy-to-use application suite. Ease of use and low complexity generally equate to improved effectiveness and lower cost of operation. A high degree of integration can aid both fault management and proactive management tasks to help deliver improved quality of service to the users of the network. Management applications and capabilities which can be built on multilayer topology information include the following:

- Visualization of the network
- Fault isolation and troubleshooting
- VLAN configuration and administration

- Traffic flow analysis and management
- Third-party applications

Visualization of the Network

By toggling between the physical, logical, and virtual network views, the network administrator can zoom in and out to navigate the network and develop a complete picture of its structure and operation. Examining the logical and physical views at the same time allows the network administrator to derive a much better understanding of the interrelationships between network elements. For example, the physical view can include the slot and port numbers corresponding to each network connection, as well as highlight the existence of parallel connections which would be obscured in a purely logical view.

The full physical view enables an analysis of the impact of link and device failures. Early identification of single points of failure which could disrupt a large number of users enables the manager to improve the survivability of the network before outages occur. The physical topology map also enables a number of other applications and serves as a convenient launch point for the remainder of the application suite, including applications which provide physical views of device front and back panels coupled with VLAN port overlays.

Fault Isolation and Troubleshooting

Visibility of the network down to the port level of hubs and switches allows a much more granular resolution of faults in the network. For example, the entire end-to-end path between the client at a remote site and the server at the central site can be traced and verified. Zooming in on the various legs of the path is aided by the topology map, making it a straightforward task to identify the failed link no matter where it is in the network. In the event a failed link has simply degraded performance, data from performance diagnostics along the alternate path can be readily accessed to identify congested links and help resolve the causes of excessive user response times.

VLAN Configuration and Administration

The virtual view of the multilayer topology database eases graphical applications for configuring VLANs which can span frame-switched, cell-switched, and configuration-switched shared-media networking technologies. For example, consider the task of moving a user connected to a configuration-switched hub from VLAN C to VLAN B. The mapping between the physical topology and the virtual network view enables the VLAN configuration application to feature a simple drag-and-drop port-level reassignment to allow the user (with ID, C3) to become designated as user with new ID, B1. This change requires new link assignments in the hub and in the switch.

Again, the physical-to-virtual mapping of the multilayer topology enables a systemwide autoconfiguration of appropriate interswitch links. Performing this sort of reconfiguration on a device-by-device basis rather than a systemwide basis would involve a large number of manual operations and careful recordkeeping to ensure that the proper mappings were made.

Traffic Flow Analysis and Management

RMON1 and RMON2 probes and embedded agents provide a wealth of real-time and historical traffic statistics at the MAC, network, and application layers. To ensure that RMON1- and RMON2-based applications realize their full potential, it is necessary to be able to map traffic data directly to both the physical and virtual views of the network.

For example, correlating RMON1 data with the physical topology is the only way to gain a complete understanding of the granularity of traffic flows through the meshed network. Network administrators can see where probes are needed and make accurate predictions about the effects of reconfiguration because they can see the relationships among all the links.

In a further example, the network administrator may want to optimize the traffic flows within a virtualized network by verifying that VLAN memberships have been appropriately assigned through analysis of inter- and intra-VLAN traffic statistics. The multilayer topology identifies the internal physical structure and the physical boundaries of the VLAN, allowing the identification of the appropriate switch ports and links from which to gather and analyze RMON1 and RMON2 data. With this information, the network administrator can determine both the node and the application which are responsible for any excessive inter-VLAN traffic, as well as identify any nodes which generate too little intra-VLAN traffic to justify membership. If only a logical or layer 3 topology of the network were available, the lack of link visibility would greatly complicate the process of monitoring intra-VLAN traffic.

RMON analysis tools should include a layer 3/layer 2 traffic matrix application capable of seeing beyond routers to allow end-to-end traffic monitoring at both the network and data-link layers. A trending application analyzes historical statistics for all LAN segments in the topology database and automatically develops baseline levels to aid the network administrator in setting threshold levels for alarms.

Third-Party Applications

Many network management tools require access to a representation of network topology. Among these are cabling management systems, inventory applications, computer-aided network cabling tools, performance management tools, and network design and simulation tools. Without automated access to a common physical topology database, network administrators using these third-party applications are forced to manually input the topology for each tool they use.

A multilayer topology database with well-documented application programming interfaces (APIs) or a standard data interchange format can make it feasible for the network administrator to readily take advantage of multiple applications which draw on topology data.

Challenges

Managing switched internetworks and VLANs pose some major challenges for network management and network administrators—and VLAN vendors, each of which currently provides only incomplete, partial management of multilayer topologies and varying degrees of integration of their own and third-party tools for their proprietary VLANs. Your choice of switch vendor determines the richness (or paucity) of the vendor's multilayer topology management solution. Meeting these challenges requires a new generation of management applications that are aware of the full complexity of network connectivity. The key enabler of these applications is a multilayer topology database which integrates and unifies the physical, data-link, network, and virtual views of the network. Unfortunately, only a few switch vendors (with hefty R&D budgets) are beginning to take the first steps toward providing total end-to-end visibility of a switched VLAN environment.

CHAPTER **5**

Standards

With so many proprietary VLAN ideas and implementations around (see Chap. 9 for details), it is essential that a standard be set up. The body to take it on is obviously the IEEE, given that so far two VLAN standards have been proposed for use, one based on IEEE 802.10 and the other on IEEE 802.1Q A standard is already set for Emulated LANs under ATM, by the ATM Forum. Other standards, notable from the IETF also have a bearing on VLANS, and are described below, beginning with spanning tree, IEEE 802.1d, the basic standard controlling the behavior of bridges (and, by implication, switches), and continuing with 802.1p, which adds important filtering controls with VLANs in mind. In this section, the major portion is devoted to the standards with the most direct effect on VLANs: 802.1Q and 802.1p.

A description of network management standards is covered in the network management section in Chap. 6. Security standards are dealt with in Chap. 7. Comments have already been made in Chap. 1 and in Chap. 4 as to the frustrations of the impinging nature of the standards from the different bodies involved and the desirability of imposing order on the current proliferation of "standard" and nonstandard VLANs.

Background

In March 1996, the IEEE 802.1 Internetworking Subcommittee completed the initial phase of investigation for developing a VLAN standard, and passed resolutions concerning three issues: the architectural approach to VLANs; a standardized format for frame tagging to communicate VLAN membership information across multiple, multivendor devices; and the future direction of VLAN standardization. The standardized format for frame tagging, in particular, known as 802.1Q, is a major step forward in enabling VLANs to be implemented using equipment from several vendors, and will encourage more rapid deployment of VLANs. In addition, establishment of a frame format specification will allow vendors to immediately incorporate the standard into their switches. All major switch vendors, including 3Com, Alantec/FORE Systems, Bay Networks, Cisco, and IBM, voted for this proposal.

However, because of the delay necessary for some vendors to incorporate the frame format specification and the drive of most organizations to have a unified VLAN management platform, VLANs will be implemented as single-vendor solutions for some time. This has significant implications for the deployment and purchase of VLANs. Unlike other LAN equipment, VLAN purchases by departments, particularly in the backbone, is impractical for organizations intending to deploy VLANs. Purchasing decisions and standardization on a specific vendor's VLAN implementation throughout the enterprise is required, resulting in a lack of cost-based product competition.

IEEE 802.1D

IEEE 802.1D is properly known as *Media Access Control (MAC) Bridges: Traffic Class Expediting and Dynamic Multicast Filtering,* but it is essentially what we know as *spanning tree.* The spanning-tree algorithm ensures the existence of a loop-free topology in networks that contain parallel bridges or switches, which are only multiport bridges, after all. A loop occurs when there are alternate routes between hosts. If there is a loop in an extended network, bridges may forward traffic indefinitely, which can result in increased traffic and degradation in network performance.

Spanning-Tree Algorithm

This problem can be avoided by implementing the spanning-tree algorithm, which produces a logical tree topology out of any arrangement of bridges. (It was designed for bridges, but, as it applies equally well to switches, this must be taken into account.) The result is that a single path exists between any two end stations on an extended network. The spanning-tree algorithm also provides a high degree of fault tolerance. It allows the network to automatically reconfigure the spanning-tree topology if there is a bridge or data-path failure.

The spanning-tree algorithm and protocol includes a procedure for notifying all bridges in the bridged local area network of topology changes and specifies a short value for the aging time which is enforced for a period after any topology change. This procedure allows the normal aging, operable during periods in which the topology does not change, to be long enough to cope with periods for which addressed end stations do not generate frames themselves, perhaps through being powered down, while not sacrificing the ability of the bridged local area network to continue to provide service after automatic reconfiguration. The aging period can be set by an administrator.

The spanning-tree algorithm requires five values to derive the spanning-tree topology. The first, a multicast address specifying all bridges on the extended network, is medium-dependent and is automatically determined by the software. You assign the remaining four values:

- Network-unique identifier for each bridge on the extended network
- Unique identifier for each bridge/LAN interface (called a *port*)
- Priority specifying the relative priority of each port
- Cost for each port

After these values have been assigned, bridges multicast and process the formatted frames [called *bridge protocol data units* (BPDUs)] to derive a single loop-free topology throughout the extended network.

In constructing a safe loop-free topology, the bridges within the extended network follow these steps:

1. Elect a root bridge. The bridge with the lowest-priority value becomes the root bridge and serves as the root of the loop-free topology. If priority values are equal, the bridge with the lowest

bridge MAC address becomes the root bridge. Each bridge designates the port that offers the lowest-cost path to the root bridge as the root port. In the event of equal path costs, the bridge examines the paths' interfaces to the root bridge. The port (interface) of the path with the lowest interface priority to the root bridge becomes the root port. If the paths' interfaces to the root bridge are also equal, then the root port is the port on the bridge with the lowest-priority value.

2. Determine path costs. The path cost is the cost of the path to the root bridge offered by each bridge port.

3. Select a root port and elect a designated bridge on each LAN. The spanning-tree algorithm selects a bridge on each LAN as the designated bridge. The root port of this bridge has the lowest-cost path to the root bridge. All bridges turn off, that is, set to blocking state, all the lines except for the single line that is the shortest-cost path to the root and any line attached to the LANs for which the bridge serves as a designated bridge.

4. Elect a designated port. The spanning-tree algorithm selects the port that connects the designated bridge to the LAN as the designated port. If there is more than one such port, the spanning-tree algorithm selects the port with the lowest priority as the designated port. This port, which carries all extended network traffic to and from the LAN, is in the forwarding state.

Thus, the spanning-tree algorithm removes all redundant ports (ports providing parallel connections) from service (places in the blocking state). If there is a topological change or a bridge or data-path failure, the algorithm derives a new spanning tree that may move some ports from the blocking to the forwarding state.

Figure 5-1 shows the resulting logical topology, which provides a loop-free topology with only a single path between any two hosts.

VLANs with Spanning Tree. A problem with spanning tree is that no matter how many paths are opened up around a network, once selected, the path set by spanning tree remains the only one to be used until a topology change or breakdown, in which case a different single path is opened. In complex switched networks with VLANs implemented, it is possible to have more than one spanning tree in operation. This will not be supported in version 1 of the 802.1Q VLAN standard (see following discussion), but some vendors have it as part of their proprietary

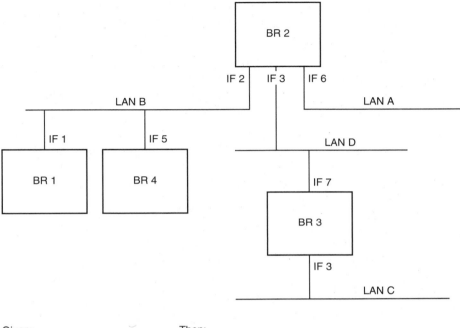

Given:

All path cards are equal.
Interface (IF) number
denotes its priority.

• • • And • • •

BR 2 – priority 1
BR 3 – priority 2
BR 4 – priority 3
BR 1 – priority 4

Then:

Bridge 1: Interface 1 is blocked.
Bridge 4: Interface 5 is blocked.

Result:

Loop-free topology
is created.

Figure 5-1   Loop-free topology

schemes and will continue to offer this as a proprietary option. The 802.1Q draft says that any given VLAN maps to a single spanning tree, but goes on to say that if a mapping choice needs to be made with multiple spanning trees, it is whether there is one spanning tree per VLAN, or whether many VLANs map to each spanning tree. Although it admits that multiple spanning trees offer some advantages over a single spanning tree in VLAN environments, it specifies only a single spanning tree since

- It may not be practical to run a spanning tree per VLAN when a large number of VLANs are configured, particularly from the point of view of scalability.

- Mapping VLANs to spanning trees may result in a more complex solution.

The IEEE does intend to review this at a future iteration of this standard.

The draft 802.1Q standard sets out how spanning trees could be used with VLANs. It points to two important items to note with respect to spanning-tree topologies:

- The spanning tree formed in a virtual LAN environment need not be identical to the topology of the VLAN(s). Each VLAN may be overlaid on different segments of the physical network topology or may be entirely separate from all the other VLANs; however, all VLANs are aligned along the spanning tree from which they are formed. In other words, a given VLAN is defined by a subset of the topology of the spanning tree on which it operates.

- The topology of the virtual LAN is dynamic. The structure of the VLAN may change because of new devices requesting or releasing the services available via the VLAN. The dynamic nature of VLANs has the advantages of flexibility and bandwidth conservation, at the cost of network management complexity.

All bridges (and switches) within a bridged LAN infrastructure participate in a single spanning tree over which multiple VLANs can coexist. As a consequence, they can be integrated into a VLAN infrastructure. VLAN-aware bridges only tag and untag frames submitted to the relay function of the bridge; frames addressed to higher-layer entities in the bridge itself are delivered directly to it and are not subject to the ingress, forwarding, and egress behavior of a VLAN-aware bridge.

IEEE 802.1p

The IEEE 802.1p *Standard for Local and Metropolitan Area Networks* is now part of 802.1D—it extends the concept of filtering services, introduced in 802.1D with a view to making it more useful in VLANs, among other characteristics. In this IEEE standard, as with spanning tree, bridges include switches, which also use the spanning-tree protocol. It defines additional capabilities in bridged LANs aimed at the provision of expedited traffic capabilities to support the transmission of time-critical information in a LAN environment, and the provision of filtering services that support the dynamic use of group MAC addresses in a LAN environment. A group associates a set of properties with a "group MAC address" that defines membership characteristics for the group, and defines the forwarding-filtering behavior of a bridge with respect to frames destined for members of that group. This is obviously necessary if dynamic addresses are to be allocated.

The protocols it comprises are

- Generic Attribute Registration Protocol (GARP)
- GARP Information Declaration (GID)
- GARP Information Propagation (GIP)
- GARP Multicast Registration Protocol (GMRP)

Bridges interconnect the separate LANs in a bridged local area network by relaying and filtering frames between the separate MACs of the bridged LAN. The aspects of service provision in bridged LANs that are altered from the standard 802.1d by 802.1p are

- Provision of the MAC service to end stations
- Preservation of the MAC service
- Maintenance of quality of service
- Provision of the internal sublayer service within the MAC bridge
- Support of the internal sublayer service by specific MAC procedures
- Filtering services

Changes are made to frame reordering, which introduces the idea of user priority, although the latter affects delay and queuing mechanisms. With 802.1p, the MAC sublayer maps the requested user priorities onto the access priorities supported by the individual MAC method. The requested user priority may be conveyed to the destination station. The transmission delay experienced by a frame in a bridge can be managed by associating user priority with the frame. Since not all MAC types are able to signal the user priority associated with a frame, MAC bridges regenerate user priority on the basis of a combination of signaled information and configuration information held in the bridge. The user priority associated with a frame can be signaled in the following ways:

- By means of the priority signaling mechanisms inherent in some MAC types (e.g., 802.4 token bus, 802.5 token ring, FDDI, 802.12 demand priority)
- By means of the user priority field carried in the tag header, as defined in P802.1Q clause 4

By this means VLANs, as defined by 802.1Q, can have a primitive quality of service based on user priority. Given the constraints placed on frame misordering in a bridge, the mappings of priority and traffic class are static. This means that once set, they cannot be changed except by manual intervention. The ability to signal user priority in the tag header (802.1Q) allows user priority to be carried with end-to-end significance across a bridged LAN, regardless of whether individual MACs that form the transmission path between source and destination are able to signal user priority.

Filtering Services

The filtering services provided in bridged LANs offer a set of capabilities that may be used to allow

- The MAC service provider to dynamically learn where the recipients of individual MAC addresses are located
- End stations that are the potential recipients of MAC frames destined for group MAC addresses to dynamically indicate to the MAC service provider which destination MAC address(es) they wish to receive
- Network managers and administrators to exercise administrative control over the extent of propagation of specific MAC addresses

Filtering services such as these increase the overall throughput of the network. They achieve this end by

- Limiting frames destined for specific MAC addresses to parts of the network which, to a high probability, lie along a path between the source MAC address and the destination MAC address
- Reducing the extent of group-addressed frames to those parts of the network which contain end stations that are legitimate recipients of that traffic

All filtering services in bridged LANs rely on the establishment of filtering rules, and subsequently the carrying out of filtering decisions, that are based on the value(s) contained in the source or destination MAC.

Bridges make use of filtering services in bridged LANs as either basic filtering services or extended filtering services. *Basic filtering services* are supported by the forwarding process and by static filtering entries and dynamic filtering entries in the filtering database. In this mode, bridges will forward all frames destined for individual and group MAC addresses for which the filtering database has no explicit filtering information in the form of static or dynamic filtering entries. Where such explicit information exists, the forwarding process will forward or filter in accordance with that information. There are two categories of basic filtering service:

1. Dynamic unicast filtering services
2. Static filtering services

Extended filtering services are supported by the forwarding process, the group registration entries in the filtering database, and the configuration of bridge filtering modes and port filtering modes. In this mode, the contents of any static or dynamic filtering entries and any group registration entries in the filtering database are taken into consideration in the forwarding-filtering decisions taken by the forwarding process. The matter in which these decisions are taken depends on the current value of the port filtering mode. There are three categories of extended filtering services:

1. Dynamic registration and deregistration services (2.6.8.1)
2. Static registration and deregistration services (clause 6)
3. Filtering mode configuration services (clause 6)

The port filtering modes are defined as follows:

- *Port filtering mode A.* "Forward all addresses" mode. In this mode, the forwarding process operates as for bridge filtering mode 1.

- *Port filtering mode B.* "Forward all unregistered addresses" submode. In this submode, the forwarding process operates as for filtering mode 1, with the additional constraint that where group registration entries exist in the filtering database, frames destined for the corresponding group MAC addresses will be forwarded only on ports identified in the member port set.

- *Port filtering mode C.* "Filter all unregistered addresses" submode. In this submode, frames destined for group MAC addresses are forwarded only if such forwarding is explicitly permitted by the information held in the filtering database.

The port filtering mode can be configured on a per transmission port basis by explicit management action and can change dynamically as a result of the operation of GMRP. The port filtering mode is ignored in bridges that are operating in basic mode.

Enhanced Internal Sublayer Service Provided within MAC Bridges

The *Enhanced Internal Sublayer Service* (E-ISS) is derived from the internal sublayer service by augmenting that specification with elements necessary in order to describe the use by the MAC bridge of the user priority information carried in tag headers (as in 802.1Q). Within the attached end station, these elements can be considered to be either below the MAC service boundary, and pertinent only to the operation of the service provider; or local matters not forming part of the peer-to-peer nature of the MAC service. Bridges that support these functions are known as "priority tag-aware bridges." The E-ISS defines the MAC service provided to the relay function in priority tag-aware bridges.

For a given MAC address, only one filtering entry may exist in the filtering database. If a dynamic filtering entry exists, then

1. Creation of a static filtering entry with the same value of MAC address will cause the dynamic filtering entry to be replaced by the static filtering entry
2. Modification of a dynamic filtering entry under explicit management control causes the entry to become a static filtering entry

Modification or removal of a static filtering entry is possible only under explicit management control.

Dynamic filtering entries are created and updated by the "learning process." They are automatically removed after a specified time has elapsed since the entry was created or last updated. This timing out of entries ensures that end stations that have been moved to a different part of the bridged LAN will not be permanently prevented from receiving frames. It also takes account of changes in the active topology of the bridged LAN which can cause end stations to appear to move from the point of view of the bridge; that is, the path to those end stations subsequently lies through a different bridge port. The aging time, after which a dynamic filtering entry is automatically removed, may be set by management. The bridge shall have the capability to use values in the range specified, with a granularity of 1 s.

GARP Multicast Registration Protocol (GMRP)

The *GARP Multicast Registration Protocol* (GMRP), part of 802.1p, provides a mechanism that allows GMRP participants to dynamically register and deregister information with the MAC bridges attached to the same LAN segment, and for that information to be disseminated across all bridges in the bridged LAN that support extended filtering services. GMRP information is of the following types:

- *Group membership information.* This indicates that one or more GMRP participants that are members of a particular group or groups exist, and carries the MAC address(es) associated with the group(s).

- *Port filtering mode information.* This indicates that one or more GMRP participants require port filtering mode A or B operation. Port filtering mode information is used to change the current port filtering mode of the port on which it is received, if the received information shows a requirement to operate in a lower port filtering mode than is currently in use on that port.

Registration of group membership information allows the bridges in a LAN to see that frames destined for the MAC address concerned are forwarded only in the direction of the registered members of the group. The group membership information registered via GMRP also allows end stations that are sources of frames destined for a group to suppress the transmission of such frames, if there are no valid recipients of those frames reachable via the LAN segment to which they are attached. This behavior on the part of end systems is known as source pruning, and it allows MAC service users that are sources of MAC frames destined for a number of groups, such as server stations or routers, to avoid unnecessary flooding of traffic on their local LAN segments in circumstances where there are no current group members in the bridged LAN that wish to receive such traffic.

IEEE 802.1Q

In March 1996, the IEEE 802.1 Internetworking Subcommittee completed the initial phase of investigation for developing a VLAN standard, and passed resolutions concerning three issues: the architectural approach to VLANs; a standardized format for frame tagging to communicate VLAN membership information across multiple, multivendor devices; and the future direction of VLAN standardization. The standardized format for frame tagging, in particular, known as 802.1Q, allows interoperability between different vendors' VLANs and will be key in encouraging more rapid deployment. Furthermore, establishment of a frame format specification will allow vendors to immediately begin incorporating this standard into their switches. All major switch vendors, including 3Com, FORE Systems (Alantec), Bay Networks, Cisco (despite its adoption of 802.10), and IBM, voted in favor of this proposal. The dynamic nature of VLANs defined by IP multicast groups enables a very high degree of flexibility and application sensitivity.

The accepted standard for VLANs will therefore be 802.1Q, with the first release due in late-1997. (Information given here on this as-yet unapproved standards draft is still subject to change.) This will provide only the most basic of facilities. Many vendors, although keen to support the standard, already offer more functionality, and will continue to do so after the release of the standard. Buyers, if not careful, will meet interoperability problems here. All will offer the standard VLAN as well as any proprietary additions. This standard is core to the future of VLANs, and is now examined in some detail.

Function

The main parts of the 802.1Q standard (as apart from existing and other draft standards covered above) have the following functions:

- Positions the function of virtual bridged LANs (VLANs) within an architectural description of the MAC sublayer
- Specifies the operation of the functions that provide frame relay in the VLAN bridge
- Defines the structure, encoding, and interpretation of the VLAN control information carried in MAC frames in a VLAN
- Specifies the rules that govern the insertion and removal of VLAN control information in MAC frames

- Establishes the requirements for, and specifies the means of, automatic configuration of VLAN topology information
- Defines the management functionality that may be provided in a VLAN bridge in order to facilitate administrative control over VLAN operation
- Specifies the requirements to be satisfied by equipment claiming conformance to this standard

Definitions

Before looking closely at this important standard, it is worthwhile to get the terms straight. So, setting out some definitions within 802.1Q:

Virtual bridged local area network—a bridged LAN in which one or more bridges are tag-aware

Tagged frame—A MAC frame that contains a tag header

Untagged frame—a MAC frame that does not contain a tag header

VLAN-tagged frame—a tagged frame whose tag header carries VLAN identification information

Priority-tagged frame—a tagged frame whose tag header carries no VLAN identification information

Tag-aware bridges—bridges that support tagging using the format defined in 802.1Q

Architecture

The architectural framework for VLANs is based on a three-level model, consisting of the following layers:

Configuration.   At this level, the concerns are how the VLAN configuration is specified in the first place, the assignment of globally significant names (ASCII names which may be propagated between bridges), and the assignment of VLAN tags.

Distribution and resolution.   This allows information to be distributed in order for bridges to be able to determine on which VLAN a given packet should be forwarded. Various possibilities exist for achieving this, including declaration protocols such as GARP and request-response protocols to request a specific VLAN association. If VLAN encapsulation is used, the distribution function will also provide mechanisms for negotiation of VLAN tags.

Mapping. At this level, the concerns are the mechanics of mapping received frames to VLANs, decisions related to where received frames should be forwarded, and mapping frames for transmission through the appropriate outbound ports (in appropriate, i.e., tagged or untagged, format).

This model is illustrated in Fig. 5-2.

The port-based approach specifies ingress (receiving), forwarding, and egress (sending) rules based on VLAN membership, which allows bridges to

- Classify all received untagged frames as belonging to a particular VLAN, as defined by the VID (virtual LAN identifier) assigned to the receiving port
- Recognize the VID associated with received tagged frames
- Make use of the VID associated with received frames in order to take appropriate forwarding-filtering decisions
- Transmit frames only through ports associated with the frame's VID

Just because it is called a "port-based approach," does not mean VLAN implementations can operate only as port-based VLANs. Rather, it defines a default, standard mode of tagging frames which have not otherwise been tagged. This is not to say that 802.1Q supports only port-based VLANs as some people have already claimed to be the case. Rather, the standard supports a default port-based classification for VLANs, implemented using the VLAN frame format specified. End stations that transmit frames in tagged format, proprietary extensions to the ingress classification rules in many bridges, and eventually bridges conforming to standardized extensions to the ingress classification rules, may actual-

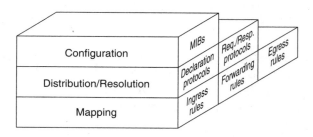

Figure 5-2
VLAN architectural framework

TABLE 5-1

Reserved VID
Values

VID Value	Meaning and Use
0	The null VLAN ID—used when the tag header contains only user priority information
1	The default PVID value used to tag frames on ingress through a bridge port

ly do much of the tagging of frames. More sophisticated tagging will be the rule for these devices, and bridges conformant to this standard will work with them. Most or all LAN segments are likely to carry tagged frames belonging to various VLANs, but each such LAN segment has its own local "default" VLAN. This "local default" defines the VLAN that untagged frames are presumed to belong to when received on the ports of 802.1Q-conformant bridges attached to that LAN segment. Table 5-1 shows the differentiation between VLAN IDs already assigned and for initial tagging on initial port entry.

VLAN technology introduces two basic types of frame tagging:

- *Implicit tagging*—a frame is classified as belonging to a particular VLAN based on the data content of the frame (e.g., MAC address, layer 3 protocol ID) and/or the receiving port

- *Explicit tagging*—a frame is classified as belonging to a particular VLAN based on an explicit VLAN tag value that is included in the frame (implying that another device such as a bridge has mapped this frame into a VLAN and has inserted the VLAN tag value)

All frames on a VLAN traversing a link must be tagged the same way on that link, either all implicitly tagged or all carrying the same explicit tag. There can be a mix of implicitly and explicitly tagged frames on a link, but only for different VLANs. A bridge adds and removes tag headers from frames, performs the associated frame translations that may be required, and forwards frames in accordance with the ingress and egress rules defined in 802.1p and also learns which VID belongs on which VLAN, updating the filtering database as it goes. A specific VID, *port VLAN identifier* (PVID), is associated with each port of the bridge and this provides the VLAN classification for frames received through that port. Also associated with each port is a *permitted frame types* parameter, which indicates whether tagged and/or untagged frames are able to be received and transmitted through that port.

IEEE 802.1Q allows each frame to belong to only one VLAN—no

overlapping VLANs can be built according to this standard. This is because each frame is assigned by a bridge and tagged according to information held by the bridge. Bridges and switches have little intelligence, and tasks performed by them need to be clearly defined.

Tagging

Tagging of MAC frames with 802.1Q headers is done to allow

- A MAC frame to carry user priority information over media types that are otherwise unable to signal priority information
- Allow a MAC frame to carry VLAN identification information
- Allow 802.5 token-ring data to be carried in native (802.5) format over non-802.5 media

Description of the tagged frame structure is based on two generic MACs which carry two generic frame formats, Ethernet and token ring/FDDI. Tagging of MAC frames requires

- The addition of a tag header to each frame. This header is inserted immediately following the destination MAC address and source MAC address (and routing, if present) fields of the original frame.
- If the source and destination media differ, tagging the frame may involve translation or encapsulation of the remainder of the frame.
- Recomputation of the *frame-check sequence* (FCS).

IEEE 802.1Q tag headers have the unfortunate effect of increasing the potential maximum frame size to over that which is allowed by Ethernet. There are several ways in which such a frame size increase could be accommodated:

- Higher-layer protocols could be used to restrict their frame sizes.
- MAC standards could change to accommodate the additional header information, while maintaining the current space limits for higher-layer protocols.
- The flexibility of the MAC specifications could be used advantageously.
- Frames could be fragmented and then reassembled.

As yet the 802.1Q committee has come up with no definitive answers to this problem, but other committees within the IEEE 802 group are cur-

Figure 5-3
Ethernet-encoded
tag header

	Octet
Ethernet-encoded TPID	1 2
TCI	3 4

rently discussing changing the standards to allow an increase in maximum frame size by several octets. This would mean a change to some of our most basic networking standards, with yet more interoperability problems associated with this.

The tag header carries the following information:

- *Tag protocol identifier* (TPID), which identifies the frame as a tagged frame

- *Tag control information* (TCI), comprising the following elements: user_priority, TR-encap, and VLAN identifier (VID)

The primary purpose of the TR-encap flag is to permit tunneling of tagged frames across a VLAN-aware region of the bridged LAN between two 802.5 LAN segments.

Structure of the Tag Header. The tag header consists of two components: the TPID and the TCI. There are two forms of the tag header, depending on the type of encoding used for the TPID. The overall structure of the two forms of header is illustrated in Figs. 5-3 and 5-4.

The Ethernet-encoded form of the tag header is used where the tagged frame is to be transmitted on 802.3/Ethernet media. The SNAP-

Figure 5-4
SNAP-encoded tag
header

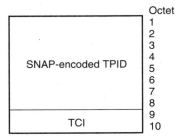

	Octet
SNAP-encoded TPID	1 2 3 4 5 6 7 8
TCI	9 10

Figure 5-5
Tag control information (TCI) format

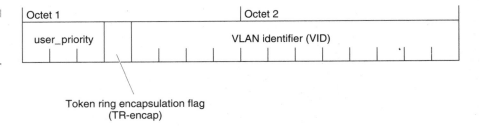

encoded form of the tag header is used where the tagged frame is to be transmitted on token ring/FDDI media.

The structure of the TPID field takes two forms, depending on whether the field is Ethernet or SNAP. The Ethernet-encoded TPID field is 2 octets in length, and carries an Ethernet V2 protocol type (TagType); the SNAP-encoded TPID (STPID) is 8 octets in length, encoded in SNAP format.

The TCI field is 2 octets in length, and contains user_priority, TR-encap, and VID fields. Figure 5-5 illustrates the structure of the TCI field.

The 12-bit VID field uniquely identify the VLAN to which the frame belongs and is encoded as an unsigned binary number.

Interoperability

Two primary aspects of the configuration of a bridged LAN are of concern in terms of interoperability:

1. Establishing a consistent view of the static filtering configuration of bridges in the bridged LAN

2. Ensuring that untagged frames are tagged (and subsequently untagged) consistently, regardless of spanning-tree reconfigurations

To ensure that end-station—end-station connectivity (or the lack of it) is consistent in all possible spanning-tree configurations, any static filters need to be established taking account of the full-mesh topology of the physical interconnections between bridges in the bridged LAN, not just the normal spanning-tree topology to which the network is configured when all bridges and LAN segments are operating correctly. An example of the consequences of failure to establish consistent controls for static VLAN filtering is that, if an access port accidentally becomes the spanning-tree-active connection between two portions of the bridged LAN, it is possible for all VLANs except one to be partitioned at that point in the topology.

Mixed Bridge Types in Networks

In networks in which 802.1Q-conformant (i.e., VLAN-aware) bridges are intermixed with VLAN-unaware bridges conformant with the 802.1D standard (spanning tree; see discussion above), the VLAN filtering services are not universally available throughout the bridged LAN; nor are tagging and untagging services. Also spanning-tree reconfigurations may cause filtering services, as well as tagging and untagging services, to become available or become unavailable independent of actions of affected users.

VLAN Membership Resolution Protocol

The proposed *VLAN Membership Resolution Protocol* (VMRP), provides a mechanism for dynamic maintenance of the contents of the port egress lists for each port of a bridge, and for propagating the information they contain to other bridges. The protocol makes use of *GARP Information Transport* (GIT), the common transport mechanism defined for use in GARP-based applications, as defined in P802.1p (see discussion above). VMRP operation is closely similar to the operation of GARP for its original purpose: specifically, registering group membership information. The main differences are

- The information carried by the protocol is 12-bit VIDs rather than 48-bit MAC addresses.
- The act of registering or deregistering a VID affects the port egress list(s) rather than the filtering database.

As with group registration, VMRP allows both end stations and bridges in a bridged LAN to issue and revoke declarations relating to membership of VLANs. The effect of such a declaration is that the VID carried in the declaration is added to the port egress list of each bridge port that receives the declaration. Once all participants on a segment that had an interest in a given VID have revoked their declarations, that VID is removed from the port egress lists.

The VMRP Entity consists of a number of components:

■ The VLAN Membership Resolution Application

■ VMRP Information Propagation

■ The GARP Entity

The details of the PDU (protocol data unit) for VRMP have not yet been fully set, but the primary features will be

■ A header consisting of a protocol identifier (identifying the PDU as a GARP PDU)

■ GARP application identifier (identifying the VMRP application as the recipient of the PDU)

■ The remainder of the PDU consists of a number of attribute records, each of which carry an operator (e.g., Join, Leave, Leave All) and any attributes that apply to the operator.

All Join.indication and Leave.indication primitives generated by the GARP entity are submitted to the VLAN Membership Resolution Application. On receipt of a Join or Leave from GARP, the Membership Resolution Application adds or removes the VID carried in the indication to the port egress list for the port on which the indication was received.

IEEE 802.10

The 802.10 committee of the IEEE has brought out a tagging format for security purposes, but some vendors have used this as a tagging type for their VLANs. In 1995, Cisco Systems proposed the use of IEEE 802.10 for VLANs. Cisco attempted to take the optional 802.10 frame header format and reuse it to convey VLAN frame tagging information instead of security information. This originated from the best of motives—vendors who foresaw early that virtual networking would be needed began to work on this before the 802.1Q committee sprang into life and the new proto-VLAN standard was not around. These vendors now have an investment in their own schemes that cannot be ignored but, even so, they have mostly announced a migration toward 802.1Q. Those who get out of step in the datanetworking world face a tough struggle.

While the 802.1Q standard uses explicit and implicit tagging that adds 2 extra bytes to every header, 802.10 uses two-level explicit tagging with a 4-byte field in the frame. Many vendors are simply not worrying about the extra 4 bytes as long as their MAC chips can send and receive packets up to 1522 bytes in length. The IEEE is proposing officially raising the packet limit for segments between switches (but not to end stations) specifically for this reason. The 1518 rule is arbitrary; sending a packet of 1519 bytes will not break anything except things which have been intentionally designed to break at more than 1518 bytes. The tag adds a VLAN ID and encapsulation header that keeps the checksum intact as it goes through, using the Said Field of the tag. IEEE 802.10 avoids a problem that occurs with 802.1Q VLANs: header size. Otherwise, there is no advantage in it, and it may in fact cause problems. For example, where 802.10 is used for VLANs, there can be problems of interoperability—where a bridge in an 802.10 VLAN is expected to use 802.10 for its security purpose, on meeting an 802.10 header, it will assume that it is part of a VLAN header and attempt to treat it as such. Thus there are 48 bytes to every packet, but this causes no performance problems as it is usually ASIC-handled, and no problems with giant frames.

For VLANs, the 802.10 tags are added as the frame reaches the first bridge (switch) port and stripped off as it leaves the last, not at the end station (an option in 802.1Q). However, it makes sense for the end stations to understand the VLAN tags, especially if multiple VLANs are involved, since otherwise they may read tagged items that they should not read. Routers can also tag and strip tags, unlike 802.1Q, setting up a subnet-to-tag association. While NetWare can route packets, if working with routers in an 802.10 VLAN, it is best to turn off this routing function and leave it to the VLAN-configured router.

Although this can be made to work technically, most members of the 802 committee have been strongly opposed to using one standard for two discrete purposes. In addition, this solution would be based on variable-length fields, which make implementation of ASIC-based frame processing more difficult and thus slower and/or more expensive.

LANE

The ATM Forum's LAN Emulation V1.0 essentially provides for an emulation of an 802.3 (Ethernet) or 802.5 (token-ring) network. FDDI is also supported. LANE provides a layer 2 switching service which is mostly transparent to existing 802 networks and network stacks above the data-link layer. One of the main purposes of LANE V1.0 is to allow ATM to be used in backbone campus networks to interconnect existing legacy

Figure 5-6
LAN emulation
(Source: Datapro
Information Services)

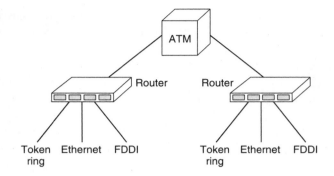

networks, that is, provide a means for LANE-compatible layer 2 bridges and switches to take advantage of high-speed interconnection, transparent to the end systems, allowing every end system to benefit from the higher aggregate bandwidth. LANE systems include the LAN Emulation Client (LEC), LAN Emulation Server (LES), and Broadcast Unknown Server (BUS). ATM LANE V1.0 provides for "best effort" service only. Figure 5-6 illustrates how LANE operates.

The ATM side of the network "emulates" the characteristics of broadcast LANs and provides MAC-to-ATM address resolution.

1. The LAN switch receives a frame from an Ethernet-connected end station. This frame is destined for another Ethernet end station across the ATM backbone. The LEC (which in this situation resides in the LAN switch) sends a MAC-to-ATM address resolution request to the LES (which in this case resides in an ATM switch).

2. The LES sends a multicast to all other LECs in the network.

3. Only the LEC that has the destination (MAC) address in its tables responds to the LES.

4. The LES then broadcasts this response to all other LECs.

5. The original LEC recognizes this response, learns the ATM address of the destination switch, and sets up a switched virtual circuit (SVC) to transport the frame via ATM cells as per AAL5, which governs segmentation and reassembly.

In looking at the path of traffic between an Ethernet-attached client and an ATM-attached server, the section that is governed by LANE extends from the LEC in the ATM interface of the LAN switch to the LEC residing in the server's ATM network interface card (NIC). From the standpoint of either MAC driver, frames pass directly between them just as if they were connected by a non-ATM backbone, with each LEC acting as a proxy MAC address. VLANs defined by port group would treat the ATM interface on the LAN switch as just another Ethernet port, and all ATM-attached devices would then be members of that VLAN. In this way, VLANs could be deployed without regard to whether the ATM switches in the backbone are from the same vendor (as long as they support LANE).

LANE can also allow for multiple ELANs by establishing more than one LEC in the ATM interfaces of participating devices (as well as a separate LES for each ELAN). Each LEC in the ATM interface of the LAN switch is treated as a separate logical Ethernet port, and each LEC in a single ATM-attached device is seen as a separate Ethernet-attached end station. Therefore, multiple LECs in a single ATM-attached device can be members of different VLANs, allowing these VLANs to overlap at ATM-attached devices. Since LANE supports only ATM-attached devices, while VLANs are defined for both ATM and non-ATM network devices, VLANs can be seen as supersets of ELANs (see Fig. 5-7).

With this structure, an ATM backbone can enable all end stations from multiple VLANs to access a centralized server or servers without passing through a router by establishing a separate ELAN for each VLAN. Since most traffic in a network is between client and server,

Figure 5-7
VLANs working with
ELANs

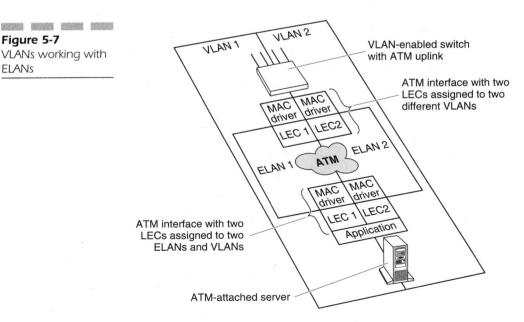

establishing VLANs that overlap at ATM-attached servers greatly reduces the number of packets that must be routed between VLANs. Of course, a small amount of inter-VLAN traffic will likely remain. Therefore, a router is still required for traffic to pass from one VLAN to another (and, therefore, from one ELAN to another).

LANE has been criticized for its CPU overhead and inefficiency when compared with other LAN technologies. This criticism is based on ATM's need to split the MTU up into 53-byte chunks. There are also interoperability problems.

MPOA

There is at least one development that may eventually standardize the route server approach to LANs within ATM. The *Multiprotocol over ATM* (MPOA) standards working group of the ATM Forum is currently working out the details of an implementation model for MPOA service. Although called "Multiprotocol," in fact the only protocol at present to be catered for within ATM is to be IP. Although various models have been proposed, MPOA is expected to provide direct virtual circuit connectivity between ATM-network-attached devices that may belong to different routing subnets. In other words, MPOA can let logical end stations that are part of different ELANs communicate directly across an ATM network without requiring an intervening router. So far, there is viable, although prestandard, MPOA from very few vendors. What there is, however, is IP switching (also Fast IP and IP Navigator), where the same method is employed. Here, an ATM (or, more rarely, FDDI) switch looks at the first packet and sends all the others in the stream after it along the same route. (Tag switching, proprietary to Cisco, does a similar job but is router-based.) This is essentially how MPOA works, and will enable the integration of IP-based networks into ATM-based networks.

Since ELANs are subsets of VLANs, MPOA holds the promise of enabling an ATM backbone to connect VLANs without the need for an external router. MPOA can be considered an enhancement beyond LANE that integrates routing functionality into the LAN-ATM edge switch. All inter-VLAN traffic would be able to leverage this capability, and network latency would be reduced.

Finalization of an MPOA standard is not expected until late 1997. It should be noted that some of the disadvantages of the route server approach, such as cost and management complexity, would remain in MPOA solutions.

Classic IP over ATM

Increasingly, VLANs are based on IP subnets, and these must interoperate with ELANs, in the future, if not now. Classic IP over ATM may be the mechanism for this, and a brief description is therefore given here.

An emulated LAN allows the ATM network to mimic a traditional LAN and thus acts as a gateway for traditional LAN end stations, which are totally unaware of the existence of ATM. An alternative is to make end stations "ATM-aware" by supplying them with an appropriate protocol stack. This stack calls on ATM address resolution services explicitly, rather than use traditional LAN broadcast-response mechanisms. A standard for such a protocol is defined by the IETF and enshrined in RFC1577. An end station using *Classic Internet Protocol* (CIP) over ATM registers its own address with an address server in the ATM network. It then uses the server to learn the ATM addresses of other CIP over ATM stations. This differs from traditional IP, whose *Address Resolution Protocol* (ARP) uses broadcasts to learn remote addresses.

CIP uses ATM's high-speed ability to provide for better "wires" between IP members. The classic logical IP subnet (LIS) is implemented over an ATM switching network using an ATMARP server to replace the broadcast ARP service. IP over ATM is a layer 3 switching service, in which IP and ARP datagrams are encapsulated in AAL5 using IETF RFC1483 LLC/SNAP encapsulation as the default. RFC1577 (IP over ATM) provides for "best effort" service only; however, work is proceeding in the IETF for mapping the Resource Reservation Protocol (RSVP) over ATM. When in place, this will enhance CIP for supporting RSVP signaling for obtaining differentiated QoS over an ATM network.

The advantages of CIP versus IP over LANE are the larger MTU size and the reduced complexity of the server. Another advantage related to performance is the delay inherent in setting up a direct SVC between IP hosts on the same LIS via IP over LANE when compared with CIP. Using LANE, the IP host must first issue an IP ARP broadcast over the LANE BUS virtual circuit to resolve the IP address to a MAC address. Then, armed with the MAC address of the destination IP host, the source must issue a LAN Emulation Client (LEC) address request over the LANE (LES) virtual circuit to resolve the MAC address to an ATM address. Only then can an ATM SVC be established. Using CIP, the IP host can directly resolve the IP address to an ATM address via the ARP server virtual circuit. (This presumes a pure IP network over ATM.) For multiprotocol support, only LANE is available and the longer SVC setup delay must be tolerated. Thus users wishing to minimize changes to their networking software might choose ELANs while users wishing to maximize performance might choose CIP.

Despite its elegance, CIP over ATM is a short-term fix. While the current version of IP, IPv4, has a QoS request mechanism called RSVP, this only requests bandwidth resources from routers. The better solution is IP version 6 (IPv6) (see section on IPv6 later in this chapter), which is gradually filtering through the IP vendor community—and is intended to be used over connection-oriented networks like ATM.

DHCP and BOOTP

As IP subnets become dominant in VLANs, IP addresses are becoming increasingly dynamically allocated, which could make it more difficult to run VLANs that rely on IP addresses as identifiers. There are three types of server-managed IP address allocation-management:

- *Manual allocation.* The server's administrator creates a configuration for the server that includes the MAC address and IP address of each client that will be able to get an address: (This is applicable to both DHCP and BOOTP, although the protocols are incompatible).

- *Automatic allocation.* The server's administrator creates a configuration for the server that includes only IP addresses, which it gives out to clients. An IP address, once associated with a MAC address, is permanently associated with it until the server's administrator intervenes.

- *Dynamic allocation.* Like automatic allocation except that the server will track leases and give IP addresses whose lease has expired to other DHCP clients

An understanding of the working of the protocols used for address allocation is necessary for anyone planning to use VLANs based on IP addresses, since this type of VLAN will not work with dynamic allocation. The following definition is from the IETF, which is the controlling body of these draft standards.

> The Dynamic Host Configuration Protocol (DHCP) provides a framework for passing configuration information to hosts on a TCP/IP network. DHCP is based on the Bootstrap Protocol (BOOTP), adding the capability of automatic allocation of reusable network addresses and additional configuration options. DHCP captures the behavior of BOOTP relay agents, and DHCP participants can interoperate with BOOTP participants. The Dynamic Host Configuration Protocol (DHCP) is built on a client-server model, where designated DHCP server hosts allocate network addresses and deliver configuration parameters to dynamically configured hosts.

The status of these standards as of March 1996 was that DHCP is an Internet Proposed Standard Protocol and is "Elective." BOOTP is an Internet Draft Standard Protocol and is "Recommended."

DHCP's purpose is to enable individual computers on an IP network to extract their configurations from a server (the "DHCP server") or servers, in particular, servers that have no exact information about the individual computers until they request the information. The overall purpose of this is to reduce the work necessary to administer a large IP network. The most significant piece of information distributed in this manner is the IP address. DHCP is based on BOOTP and maintains some backward compatibility. The main difference is that BOOTP was designed for manual preconfiguration of the host information in a server database, while DHCP allows for dynamic allocation of network addresses and configurations to newly attached hosts. Additionally, DHCP allows for recovery and reallocation of network addresses through a leasing mechanism.

A DHCP *lease* is the amount of time that the DHCP server grants to the DHCP client permission to use a particular IP address. A typical server allows its administrator to set the lease time. If there are more users than addresses on the network, the lease time should be kept short so that people don't end up sitting on leases. Naturally, there are degrees. In this situation, this could be 15 min, 2 h, and 2 days. On the other hand, if a typical user is on for an hour at minimum, that suggests an hour lease at minimum. While some sites would manually allocate any address that people expected to remain stable, other sites would want to use DHCP's ability to automate distribution of relatively permanent addresses. A very relevant factor is that the client starts trying to renew the lease when it is half way through.

Having a server administrate DHCP resolves issues of security, IP services, and compatibility:

- A computer that needs a permanently assigned IP number might be turned off and lose its number to a machine coming up. This has problems both for finding services and for security.

- A network might be temporarily divided into two noncommunicating networks while a network component is not functioning. During this time, two different client machines might end up claiming the same IP number. When the network comes back, they would start malfunctioning.

- If such dynamic assignment is to be confined to ranges of IP addresses, then the ranges are configured in each desktop machine rather than being centrally administered. This can lead both to hidden configuration errors and to difficulty in changing the range. Another problem with the use of such ranges is keeping it easy to move a computer from one subnet to another.

In a subnetted environment, the DHCP server discovers what subnet a request has come from by means of DHCP client messages, sent to off-net servers by DHCP relay agents, which are often a part of an IP router. The DHCP relay agent records the subnet from which the message was received in the DHCP message header for use by the DHCP server. (*Note:* A DHCP relay agent is the same thing as a BOOTP relay agent, and technically speaking, the latter phrase is correct.)

A single LAN might have more than one subnet number applicable to the same set of ports (broadcast domain). Typically, one subnet is designated as primary; the others, as secondary. A site may find it necessary to support addresses on more than one subnet number associated with a single interface. DHCP's scheme for handling this is that the server has to be configured with the necessary information and has to support such configuration and allocation.

If a physical LAN has more than one logical subnet, how can different groups of clients be allocated addresses on different subnets? One way to do this is to preconfigure each client with information about what group it belongs to. A DHCP feature designed for this is the user class option. To do this, the client software must allow the user class option to be preconfigured and the server software must support its use to control which pool a client's address is allocated from.

ARP and ICMP

An entity on a network generally has two addresses the burned-in MAC address and the network layer or IP address, which is usually configured in the protocol stack software, and is used to convey the message from logical endpoint to logical endpoint. Ethernet and token-ring cards only know how to recognize the MAC addresses; therefore, any communicator on a network must be able to specify the MAC address of both itself and the destination of its messages. Just knowing that you want to talk to a particular IP address is not sufficient.

Furthermore, a message traveling from one logical endpoint to another logical endpoint (station to station, for example) might in reality pass through several different NICs before reaching its destination. In the TCP/IP world, a packet destined for a station that is several gateways away would be addressed to the destination station at the network layer, but at the MAC layer would first be addressed to the first gateway, then the second, then the third, and so on, until the final gateway addressed the packet to the destination.

Two methods are used for address resolution in IP networks. The first method, used in IP version 4, uses ARP (Address Resolution Protocol); this method has been used since the 1970s and is considered the standard. ARP for Ethernet is defined by IETF RFC826. There are other specifications for ARP over FDDI, HIPPI, ATM, and other media. The new method, in IP version 6 (IPng, "IP Next Generation," the final drafts of which were released in January 1995) uses ICMP (Internet Control Message Protocol) for address resolution. ICMP has been in use as a protocol since the 1970s, but its use in address resolution is brand-new with IP version 6. In 1997, the deployment of IPng (version 6) is limited and it will be a number of months (at the time of writing) before the implications of ICMP resolution become significant to most networks. ARP is used on IEEE 802 networks to find the 48-bit MAC address associated with a given IP address. ARP packets are not the same type of packets as used in IP—they contain protocol type fields so that ARP can be used by other network protocols which need to translate their addresses to media-specific addresses. ARP requests are necessarily sent as broadcasts; replies are sent unicast.

A router or a layer 3 switch uses its routing table to determine the destination for a frame (unless it is not the same subnet, in which case the frame is delivered without consulting the table), which will be either a remote IP address or the awareness that the frame should be delivered directly.

An ARP frame has four significant fields relative to this discussion:

- *Sender's hardware address*—the data-link address of the sender
- *Sender's protocol address*—the IP address of the sender
- *Target hardware address*—the data-link address of the target being sought
- *Target protocol address*—the IP address of the target being sought

Not all four of these fields are filled in by the sender of an ARP frame; unknowns are left as zeros. This ARP command frame gets to the Ethernet or token-ring broadcast destination as everyone anywhere on the network hears the frame. When the destination node hears the frame, it responds with an ARP reply, supplying its MAC address, thus linking the IP address and the MAC address. The association between IP address and data-link address is recorded in a temporary memory table referred to as the *ARP cache*. The ARP cache tells a station what data-link destination address to use for every IP destination address.

When a station first boots up, it must ARP for the IP address of the default gateway. Additionally, prior to sending to any destination IP address, there must be an entry in the ARP cache associated with that destination IP address. The ARP cache is dynamic; entries age out after just a few seconds if they are unused. Consequently, during a typical conversation a station ARPUs for the destination once at the beginning and then the entry remains in the ARP cache for the life of the connection (assuming that the connection is active).

The two tables (the routing table and the ARP cache) form the basis for frame transmission and forwarding. The address mask serves as the guide for interpreting the IP address. The default gateway is the destination specified in the routing table to use when no other destination can be ascertained. The idea of a default destination has no corresponding behavior in the ARP cache.

Some vendors implement a special feature called *proxy ARP* on their routers. When a host on one network segment broadcasts an ARP request for an IP address which belongs to some other net or subnet, any router which believes itself to be the next hop to the target destination will send a reply which claims that its own MAC address belongs to the target. The sending host will then direct traffic for the target to such a router, which is generally what it should do.

Classic IP and ARP over ATM

Where an ATM environment exists as a logical IP subnetwork (LIS), the operation of ARP is straightforward. The ARP request basically says, as normal, "Will the station on this Ethernet channel that has the IP address of 192.0.2.2 please tell me what the address of its Ethernet interface is?" A given Ethernet system can carry several different kinds of high-level protocol data. For example, a single Ethernet can carry data between computers in the form of TCP/IP protocols as well as Novell or AppleTalk protocols. The Ethernet is simply a trunking system that carries packages of data between computers; it doesn't care what is inside the packages.

IPv6

IP, regulated and developed by the IETF, is showing the strain as today's networks seek to interconnect the world's billions of people. Not only is the Internet growing in popularity, leading to many more IP addresses being needed, but also corporate accounts are switching to pure TCP/IP networks, with all connected computers needing to have IP addresses of their own. IP, now at version 4, cannot offer enough addresses for the billions of people wanting them. Addresses are not the only issue— again, the Internet's popularity is leading to "Internet meltdown," in which constant congestion (known as "lag") is the bane of life to those connected. While the necessary increase in bandwidth is promised, it is bound to be used up in an increase of multimedia data transmitted. (At present the Internet is mainly text- and graphics-based.)

The next generation of the Internet Protocol (IPv6) is intended to support Internet traffic for many years into the future by providing enhancements over the capabilities of the existing IPv4 service. IPv6 (also known as IPng, for "new generation" will offer a vastly expanded addressing scheme, security enhancements, and other features that will facilitate truly global interconnection.

This summary of the IPv6 standard, compared with the already well-known IPv4, is here because it is relevant in any discussion that involves DHCP and any kind of encapsulation and tunneling, all of which are vitally different under the new standard and its extensions, one of which follows the description of IPv6.

IPv4

Each TCP/IP network interface requires a unique IP address, which IP routers use to forward packets as needed across the various network cables that connect communicating systems. An IPv4 address is 32 bits long (see Fig. 5-8).

IPv4 recognizes three kinds of unicast (that identifies a specific network interface) addresses:

- *Class A addresses,* which consist of a 7-bit network number, followed by a 24-bit host number. These addresses are intended for use by the world's 128 largest organizations. Class A addresses are all already assigned.

- *Class B addresses,* which consist of a 14-bit network number, followed by a 16-bit host number. As many as 16,384 organizations can claim class B addresses, supporting communities with up to 65,536 members. Class B addresses are also practically all allocated.

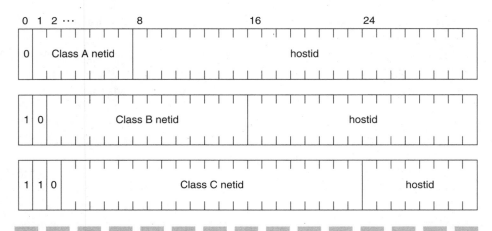

Figure 5-8 IPv4's 32-bit address field, showing its three unicast address formats (Source: IETF)

■ *Class C addresses,* which consist of 21-bit network numbers. Class C addresses are intended for small network communities with 256 connections or less. From a pool of a little more than 2 million possible values, growth in today's Internet is being sustained by assigning blocks of unused class C addresses to new subscribers, as well as by creatively reusing certain previously assigned addresses.

IPv4 also recognizes a fourth kind of address: class D, not assigned to a specific end station. Instead, it identifies the members of a logical group of interfaces. A packet intended for a class D address will then be delivered to all of its logical holders. This multicast addressing scheme was never widely used.

Following the IPv4 rules as they were originally conceived, any medium-sized company with more than 256 computers would apply for a class B address, and consequently tie up 64,000 values, mostly remaining unused in many cases. Large companies claiming class A addresses tie up (and also grossly underutilize) about 16 million of IPv4's available addresses. The introduction of CIDR (Common Interdomain Routing Protocol), which (among other things) permits the assignment of class C addresses in consecutive blocks (aggregates) to build mini—class B networks, has bought some time for the IPv4 address pool.

Classifying millions of computers with just two hierarchical levels results in very large routing tables—incurring enormous processing overhead—in the Internet's interior routers. If the original IPv4 architecture were in place today, the Internet's routers would have to maintain millions of paths to class C networks. Again, the CIDR address aggregation scheme flattened the growth of the Internet's routing tables, and bought more time for IPv4.

To preserve the world's enormous investment in TCP/IP, IPv6 can coexist indefinitely with IPv4 in both host computers and routers. As IPv6 implementations become available, systems managers and network administrators need to upgrade only as many devices at a time as they can handle, over as long a period of time as necessary. While the IPv4-to-IPv6 details for any particular device will depend on the vendor's implementation, the promise of gradual and (relatively) painless migration rests on several key characteristics of IPv6 architecture:

■ IPv6 is designed to work with IPv4. Programmers who want to take advantage of new IPv6 features (see section below) can port their applications gradually to the new IPv6 APIs.

■ IPv6 can run on the same machine concurrently with IPv4. Unless explicitly configured as an IPv6-only node, an IPv6 computer system will be able to communicate with its IPv4-equipped neighbors transparently.

■ IPv6 provides a feature that enables IPv6 packets to be "tunneled" through existing IPv4-routed networks. This feature will enable network managers to phase IPv6 routing equipment gradually into their networks, and to use today's IPv4 Internet as a routing path to interconnect their IPv6 sites.

IPv6 Features

Address Structure. IPv6 addresses are 128 bits long, but the increased address size is less important than the address structure. IPv6 recognizes three kinds of unicast addresses (see Fig. 5-9) offers a new multicast address format, and introduces the new anycast address:

1. Provider-based unicast addresses are assigned by an Internet Service Provider (ISP) to an organization, offering globally unique Internet addresses to all the organization's members for easy integration within the worldwide Internet community. Devised as part of CIDR, the basic mechanism for assigning these addresses through ISPs is already in place.

2. Site-local-use addresses are assigned to the network devices within an isolated intranet. If the organization later joins the Internet

Figure 5-9

IPv6's 128-bit address field, showing its three unicast address formats (Source: IETF)

3 bits	n bits	m bits	o bits	p bits	125 − (n + m + o + p) bits
010	Reg. ID	Provd. ID	Subsc. ID	Subnet ID	Intf. ID

10 bits	n bits	118 − n bits
1111111010	000...0	Intf. ID

10 bits	n bits	118 − n bits
1111111010	000...0	Intf. ID

community, its sitewide local addresses all automatically become globally unique provider-based addresses with one administrative operation. (With IPv4, all addresses have to be changed manually if DHCP is not in use.)

3. Link-local-use addresses are designed for use by individuals on a single communications link—such as are connected through phone lines (voice or ISDN) or radio links.

IPv6 will offer a "stateful" autoconfiguration approach, similar to today's DHCP but also "stateless" autoconfiguration. Especially useful for mobile computing applications, IPv6 nodes will be able to generate globally unique addresses by concatenating the link-local-address of the network connection they're using with an internal interface number, such as an Ethernet or token-ring MAC address.

With IPv6, multicast addressing is likely to find wider use. IPv6's new multicast address format allows for trillions of possible multicast group codes, each identifying two or more packet recipients. In addition, the scope of a particular multicast address can be confined to a single system, restricted within a specific site, associated with a particular network link, or distributed worldwide.

IPv6 also introduces a new kind of address: *anycast,* a single value assigned to more than one interface. A packet sent to an anycast address is routed to the nearest interface having that address.

There are no broadcast addresses in IPv6; their function has been superseded by multicast addresses.

Header Structure.   In designing the new header structure, the IETF considered two important principles:

■ A simple approach should be taken.

■ The architecture should be extendible.

For all its simplicity, even the basic form of an IPv6 header offers some important new features (see Fig. 5-10).

With its greatly enlarged payload length field, IPv6 can carry oversized messages, a feature that offers significant performance advantages for applications running over fast, reliable networks with large MTU (maximum transmission unit) characteristics. To use large datagrams, the communicating hosts will need to negotiate an MTU appropriate to the physical network paths between them, falling back to the smallest available.

In a related development, IPv6 packet segmentation will work differently from IPv4. Where IPv4 routers break a large IP packet dynamically

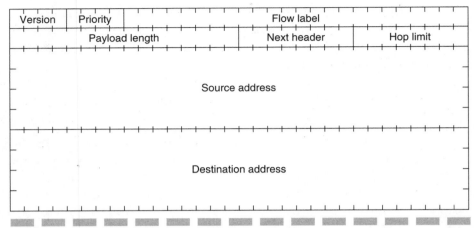

Figure 5-10 IPv6 packet header format (Source: IETF)

into smaller segments as it moves through a network, IPv6 routers will not need to. When the transition to IPv6 is complete, the Internet will consist only of networks with MTUs equal to or larger than 576 bytes, with IP packet segmentation and reassembly handled exclusively in the communicating hosts.

Tunneling. IPv6 tunneling (the draft dates from December 1996) is a technique for establishing a virtual link between two IPv6 nodes for transmitting data packets as payloads of IPv6 packets. With respect to the two nodes, this virtual link, called an *IPv6 tunnel,* appears as a point-to-point link on which IPv6 acts like a link-layer protocol. The two IPv6 nodes play specific roles. One node encapsulates original packets received from other nodes or from itself and forwards the resulting tunnel packets through the tunnel. The other node decapsulates the received tunnel packets and forwards the resulting original packets toward their destinations.

An IPv6 tunnel is a unidirectional mechanism—tunnel packet flow takes place in one direction between the IPv6 tunnel entry-point and exit-point nodes. Bidirectional tunneling is achieved by merging two unidirectional mechanisms, that is, configuring two tunnels, each in opposite direction to the other—the entry-point node of one tunnel is the exit-point node of the other tunnel.

IPv6 encapsulation consists of prepending to the original packet an IPv6 header and, optionally, a set of IPv6 extension headers which are collectively called *tunnel IPv6 headers*. The encapsulation takes place in an IPv6 tunnel entry-point node, as the result of an original packet being forwarded onto the virtual link represented by the tunnel. The original packet is processed during forwarding according to the forwarding rules of the protocol of that packet.

Quality of Service. Usually spoken of in the context of ATM, QoS is to become part of IP, too, which will open the door to multimedia in the IP world without the need to go via ATM. The features are to be flow labels and priority. A *flow* is a sequence of packets sent to a unicast or multicast destination. The IPv6 flow label enables the flow's source to identify a logical sequence of packets; routers that support this feature can then maintain a context for the flows currently in transit. This will result in optimized performance and congestion management—resulting in faster, more reliable networking. Communicating hosts can also specify a packet's priority, a feature that will allow IPv6 routers to discriminate and favorably accommodate TCP/IP applications that require faster response time.

IPv6 Security Features. In the original standards document, the IETF intended that algorithms for authentication and privacy be carried as IPng extension headers and include an initial selection of required encryption and key management algorithms plus a facility to support other optional algorithms. The IPv6 authentication header prevents unauthorized hosts from sending traffic to certain destinations by first forcing the sender to log into the receiver in a secure way. As things stand, however, this extension to the protocol leaves the specifics of the authentication algorithm up to the implementers.

To prevent the interception of sensitive traffic, the IPv6 *encapsulating security header* allows the IPv6 traffic exchanged between two hosts to be encrypted. Also algorithm-independent, its standard use of DES (Data Encryption Standard) CBC encryption will mean secure traffic even through the Internet.

The IPv6 security specifications provide for encapsulating the IP security payload in IPv4, essentially emulating the *next header payload* field of IPv6. However, while this implementation is an optional issue in IPv4, it is a mandatory part of IPv6.

The IETF is also examining IPv6 firewall issues and if necessary will develop specific firewall frameworks.

GSE for IPv6. The IPv6 standard was to be made firm by 1995, but there are still some problems to be resolved. On February 27—28, 1997, the IPng Working Group held an interim meeting in Palo Alto, California to consider adopting a new "GSE [global, site, end system]—an Alternate Addressing Architecture for IPv6" proposal to reduce the reliance on CIDR and make renumbering more palatable.

The problem is that CIDR-style provider-based aggregation breaks down in the face of the accelerating growth of multihomed sites (leaf sites or regional networks). As sites have become more dependent on the Internet, they have begun to install additional connections to the Internet to improve robustness and performance. Such sites are called *multihomed*. Unfortunately, when a site connects to the Internet at multiple places, the impact can be much like a site that switches providers but refuses to renumber.

In the pre-CIDR days, multihomed sites were typically known by only one network address. When that site's providers would announce the site's network into the global routing system, a shortest-path type of routing would occur so that pieces of the Internet closest to the first provider would use the first provider and other pieces would use the second provider. This allowed sites to deal with the load on their multiple connections with the routing system itself. With CIDR, if a multihomed site is known by a single prefix taken from one of its providers, then that prefix is aggregable by the provider which assigned the address but not aggregable by the other providers.

One way to prevent entropy from taking over under CIDR is to have multihomed sites use address space from all its providers. Although this in itself is not so difficult, it changes the way load sharing is handled, complicating the engineering by requiring much more foresight and introducing complexities such as DNS (Domain Naming System) and its caching system. So, like those sites which refuse to renumber, many multihomed sites today are known by a single prefix, thus reducing the efficiency of the global routing system.

Worse than this problem, renumbering an entire site to accomplish a simple topological rehoming such as changing ISPs is a problem whose magnitude can only grow over time. It will remain increasingly difficult to explain this renumbering requirement to customers. While the large IPv6 addresses provide for a huge increase in the number of end systems which can be accommodated, this also portends a huge increase in the number of routes required to reach them. Even if CIDR aggregation were to continue at current levels (maintaining current efficiency is relatively unlikely), this still presents a serious problem for the growth of the global route computations.

GSE for IPv6 presents a new proposal for using the 16-byte IPv6 address which mitigates the route scaling problem and with it a number of collateral issues. This model provides for aggressive topological aggregation while controlling the complexity of flat-routed regions. It exploits and supports the dynamic address assignment machinery in IPv6 but makes the exact role of that machinery a decision local to a site.

Architecture. In GSE, 16-byte IPv6 addresses are split into three portions: a globally unique *end-system designator* (ESD), a *site topology partition* (STP), and a *routing goop* (RG) portion. The STP corresponds (roughly) to a site's subnet portion of an IPv4 address, whereas the RG identifies the attachment point to the public Internet. Routers use the RG + STP portions of addresses to route packets to the link to which the destination is directly attached; the ESD is used to deliver the packet across the last hop link. An important idea in GSE is that nodes within a site would not need to know the RG portion of their addresses. Border routers residing between a site and its Internet connect point would dynamically replace the RG part of source addresses of all outgoing IP datagrams, and the RG part of destination addresses on incoming traffic.

The architecture is based on a few central concepts:

- A strong distinction between public and private topologies
- A strong distinction between system identity and location
- GSE—global, site, and end-system address elements
- The deep similarity of rehoming and multihoming
- Rewriting address prefixes at site boundaries
- Very aggressive hierarchical network topology aggregation
- Optimizing actual forwarding paths by limited-scope cut-throughs

This model draws a strong distinction between the public topology which forms the transit infrastructure of the global Internet and a site which can contain a rich but strictly private local network topology which cannot leak into the global routing machinery. The site is the fundamental unit of attachment to the global Internet and is therefore strictly a leaf (using the tree model), even if possibly multihomed. A very strong distinction is also made between the identity of a computer system and where it attaches to the public topology. In IPv4 and current IPv6 models, these notions of identity and location are deeply commingled, and this is the fundamental reason why simple topology changes have such wide-ranging impact on address assignment (if aggregation is to be maintained at all).

The main proposed changes to the IPv6 plan include

- Making changes to the IPv6 provider-based addressing document, to facilitate increased aggregation.

- Creating hard boundaries in IPv6 addresses to clearly distinguish between the portions used to identify hosts, for routing within a site, and for routing within the public Internet.

- Designating the low-order 8 bytes of IPv6 addresses to be a globally unique ESD. This change has potential benefits to future transport protocols (e.g., TCPng).

- Make a clear distinction between the "locator" part and the "identifier" part of the address. The former is used to route a packet to its endpoint; the latter is used to identify an endpoint independent of the path used to deliver the packet. Although this is a potentially revolutionary change to the IPv6 addressing model, existing transport protocols such as TCP and UDP (User Datagram Protocol) will not take advantage of the split. Future transport protocols (e.g., TCPng), however, may.

The key departure of GSE from classic IP addressing (both v4 and v6) is that rather than overloading addresses with both locator and identifier purposes, it splits the address into two elements: the high-order 8 bytes for routing and the low-order 8 bytes for unique identification of an endpoint.

Another fundamental aspect of GSE is that site border routers rewrite addresses of the packets they forward across the site/Internet boundary. In addition to rewriting source addresses on leaving a site, destination addresses are rewritten on entering a site.

GSE defines a specific mechanism for providers to use to support multihomed customers that gives those customers more reliability than singly homed sites, but without a negative impact on the scaling of global routing. This mechanism is not specific to GSE and could be applied to any multihoming scenario where a site is known by multiple prefixes (including provider-based addressing).

The most important feature of the ESD part of the proposal is its global uniqueness. Endpoints of communication would care only about the ESD, and on receipt of a datagram only the ESD would be used in testing whether a packet is intended for local delivery. An ESD that uniquely identifies an endpoint of communication (independent of the interface through which that was reached) should be globally unique so that a node that receives a packet can definitively determine whether the packet is intended for it by comparing only the ESD portion of the address.

Conclusions

All this and much more besides (the short summary above deals only with the main points) is still being discussed within the IETF. Much of GSE will probably not make it into the standard. So far, the IETF concludes that there are problems in several areas of the proposal. A few of these are

- It is not adequate to use an ESD alone for endpoint identification because of the ease of hijacking a TCP connection. Incoming packets with a wrong ESD can easily be detected as coming from an incorrect source; however, incoming packets with a correct ESD cannot be easily trusted as being from the correct source.

- There are incompatibilities between the IPv4 with ARP and the ESD, especially in the case of stations mistakenly being allocated the same IP number. At present, the system deals with the resulting confusion adequately, but here traffic may actually get through to the wrong node.

- Adding sufficient structure to an 8-byte ESD requires more bits than are compatible with stateless autoconfiguration, and required databases will probably be poorly maintained. Imposing a required hierarchical structure on ESDs would also introduce a new administrative burden and a new or expanded registry system to manage ESD space. While the procedures for assigning ESDs, which need only organizational and not topological significance, would be simpler than the procedures for managing IPv4 addresses (or DNS names), it is hard to imagine such a process being universally well received or without controversy; it seems a laudable goal to avoid the problem altogether if possible.

- Finally, there is an argument based on allocation efficiency. Since GSE uses 64 bits to designate the site and subnet and the same number to designate the end system, the allocation efficiency for the latter assignment process must be much greater.

There are also many advantages, particularly with respect to multihoming, and the redistribution of load so that those who reap the rewards also pay the most, but in short, while this scheme addresses some of the issues raised by the continued use of CIDR in IPv6, it also introduces, at a very late stage in the standards process, a whole set of new ideas. While GSE as a whole may well not be adopted, it has already caused the IETF to rethink IPv6, and large parts of it may be rewritten or have major extensions added. What is certain is that the introduction of these new ideas will hold up the final release of the IPv6 standard.

Managing VLANs

The Need for VLAN Management

Some organizations have installed high-speed and switched networks and advanced technologies, such as VLANs, to improve network performance. However, unexpected problems, delays, and inefficiencies resulting from the adoption of new technologies can be time-consuming and risky.

To reduce deployment problems and lower implementation risks associated with new technologies, a thorough understanding of the network and the impact of new technologies—top-to-bottom and end-to-end information—is required. By combining RMON and network management tools with distributed data collection and analysis consoles, a detailed picture of traffic flow patterns, network utilization, and protocol turnaround times can be constructed. (See Fig. 6-1.)

Figure 6-1
Traffic management
in an ideal world

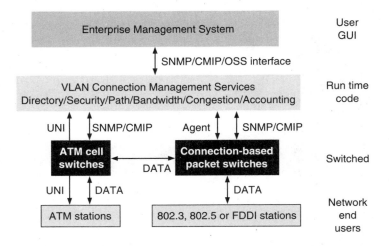

Analysis and Planning for VLANs

Having determined that VLANs need to be a part of network planning in the immediate future, we can assume that server access, server location, and application utilization must all be thoroughly analyzed to determine the nature of traffic flow in the network (see Fig. 6-2). This analysis should answer the remaining questions about where VLAN broadcast domains should be deployed, what role ATM needs to play, and where the routing function should be placed.

When installing a switched network or VLANs, concrete information, not guesswork, is needed. Specifically, these steps should be addressed:

- When planning a migration to switched networks/VLANs, learn as much as possible about the existing network.

- Determine where the critical segments, users, and applications are, together with the impact of downtime and poor performance.

- Use distributed protocol analysis and troubleshooting tools to examine all seven layers of the OSI model to gather information on application performance, top network conversations (RMON TopN, i.e., busiest device-to-device conversations), and server bottlenecks.

- On the basis of this information, determine which elements of the network need to be upgraded in order to support VLANs without degradation of response times.

- Pinpoint inefficiencies, such as excessive retransmissions, timeouts,

Figure 6-2
Monitoring switches,
networks, and
VLANs. (Source:
3Com)

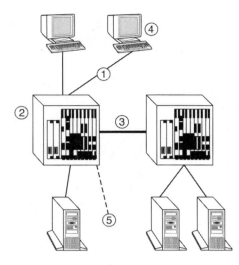

① On shared media

② In hub or switch

③ On backbone/WAN

④ At desktop

⑤ Copy to port

unexpected network hogs (RMON TopN), and poor frame sizing, in order to determine bandwidth problems.

▪ Use network modeling and capacity planning tools to create a modeling environment to baseline the current network. This means that "what if" scenarios can be simulated, enabling VLAN implementations, new technologies, or routing changes to be planned effectively.

▪ Design in fault and performance management tools which match the implementation of VLANs. If a Fast Ethernet or ATM backbone is planned, plan the purchase of Fast Ethernet or ATM RMON probes.

Traffic Management in Switched Networks

SNMP Management

Two of the network manager's most popular management allies have been the protocol analyzer and the SNMP (Simple Network Management Protocol) manager. The protocol analyzer allows the network manager to instantly go into a problem network and find and fix the problem which causes a failed or dysfunctional LAN. For many years, another popular tool to listen for a properly functioning network has been the SNMP management station. Consisting primarily of software running on a PC or UNIX workstation, the SNMP management station is able to communicate with the various devices on a network and detect possible problems which demand attention. If the network is functioning properly, it is the SNMP management station's primary task to display and log these network statistics in easy-to-interpret information management displays. SNMP was developed by the Internet Engineering Task Force (IETF) to allow for the remote management of network equipment. Embedded agents in hubs, switches, bridges, routers, and even print server equipment allow the management station to get and set desired parameters in an enterprise network. The network equipment supporting the SNMP protocol maintains a set of management information bases (MIBs) of network statistics.

The SNMP management station queries these MIBs for network data to give the network manager a picture of the relationship between the device and the network, putting together a picture of how the entire network is performing. Agents continually gather statistics, such as the number of bytes received, and record it in an MIB. The polling by the network management station is well defined in the SNMP standard.

Unsolicited responses from network devices to the SNMP management station can occur. These responses are called "traps." The trap capability allows a network element to notify the management station that an emergency condition has occurred. However, the level of support for SNMP varies from product to product. This is because SNMP makes demands on processing power. Consequently, hardware developers have to make the choice of either partially implementing SNMP or increasing the specifications and price. Thus, some products implement only a subset of SNMP features.

SNMP Architecture.   The client program (called the *network manager*) makes virtual connections to a server program (called the *SNMP agent*) executing on a remote network device. The database controlled by the SNMP agent is referred to as the SNMP MIB and is a standard set of statistical and control values. SNMP also allows the extension of these standard values with values specific to a particular agent through the use of private MIBs.

Commands, issued by the network management client to an SNMP agent, consist of the identifiers of SNMP variables (referred to as *MIB object identifiers* or *MIB variables*) along with the instructions to either get the value of the identifier or set the identifier to a new value.

Through the use of private MIB variables, SNMP agents can be tailored for a myriad of specific devices, such as network bridges, gateways, routers, and switches. The definitions of MIB variables supported by a particular agent are incorporated in descriptor files, written in *Abstract Syntax Notation* (ASN.1) format, made available to network management client programs (i.e., SNMP browsers) so that they can be aware of these MIB variables and their usage.

Network managers can obtain this information by sending queries to the agent's MIB, a process called *polling*. While MIB counters record aggregated statistics, they do not provide any historical analysis of daily traffic. To compile a comprehensive view of the day's traffic flow and rates of change, the management station must poll SNMP agents continually, every minute of every day. In this way, network managers can use SNMP to evaluate the performance of the network and to uncover traffic trends, such as which segments are nearing traffic capacity or are needlessly corrupting traffic. Advanced SNMP network management stations can even be programmed to disable ports automatically or take other corrective action in reaction to historical network data.

Disadvantages of SNMP.   Several weaknesses of SNMP can be identified. First, SNMP is not a particularly efficient protocol. Bandwidth is wasted with needless information, such as the SNMP version (transmitted in every SNMP message) and multiple length and data descriptors scattered throughout each message. Second, the method through which SNMP variables are identified (as byte strings, where each byte corresponds to a particular node in the MIB database) leads to large data handles which consume substantial parts of each SNMP message.

Another disadvantage of the SNMP specification is the protocol used to extract information from the SNMP agents. The protocol extracts information through polling all agents. SNMP polling has two distinct disadvantages:

- ■ It does not scale well. In large networks, polling generates substantial network management traffic, contributing to congestion problems.
- ■ It places the burden of collection on the network management console. Management stations which can easily collect information on 8 segments might not be able to keep up when monitoring 48 segments.

SNMP Compatibility. This problem is exacerbated by some vendors branding their products as "SNMP-compatible," even though they may support only a subset of SNMP's MIB. To this, add the fact that some products, such as hubs and switches, have no agreed-on standard MIB and that vendors are allowed to define their own "private" MIB extensions. This undermines to some extent SNMP's universality.

The result is that, say, an SNMP manager for Bay Networks is ideal for managing Bay Network switches but not quite perfect in managing Cisco switches—they use the same language, but the vocabularies are very different. Some manufacturers pride themselves on supporting the MIB extensions from others, but very often even this does not work well; companies may publish their MIBs, but do not keep their public versions up to date—to continue the analogy, it is as if an out-of-date dictionary were being used. Some of SNMP's limitations can be overcome by the use of SmartAgents in devices that can make corrections automatically.

From the previous discussion, it can be seen that the usefulness of a centralized SNMP management scheme was reaching its limits and a new form of remote management was required. While SNMP was designed to look at network devices, it was not designed to look at the network in a holistic fashion. Table 6-1 summarizes the network management products examined so far.

The IETF's response to these problems was the RMON standard.

TABLE 6-1

SNMP Device Management

Product or function	Strengths	Weaknesses
LAN analyzer	Portable	Network-oriented
	Reactive	Expensive
		Segment view
SNMP device management	Remote management	Single-vendor solution
		Device-oriented
		One device at a time
SNMP network management	Enterprise view	Device-oriented
	Remote management	Polling does not scale
		Multivendor solution

RMON

In 1991, an extension of SNMP called RMON (Remote MONitoring) protocol was created by the IETF and defined in RFC1757 (which supersedes the original RFC1271). Just as SNMP operated with the use of intelligent agents to send back information, RMON uses intelligent agents to provide filtered data and information when it is required by the SNMP management station. RMON complements to SNMP. It reduces the polling which had previously hampered the use of SNMP on larger networks and extends the range of information which can be sent back to the SNMP manager.

RMON has been widely implemented by vendors, although, surprisingly, many claim RMON compliance without including support for all groups. This is because all groups are optional and means that only one group has to be supported to claim compliance with the RMON standard. Worse, this is regarded as acceptable practice even by major players. In addition, RMON has its limitations, leading to fragmentation of the standard, as some vendors extend RMON considerably in a proprietary manner.

Another concern is the impact that the RMON protocol has on the network it is supposed to monitor. The network transport for RMON is SNMP, which generates a surprisingly high level of overhead when transporting information from agent to client; as a result, it soaks up network capacity for little reward. This is made worse by the requirement that RMON agent sends it a complete library of statistics, irrespective of whether changes have occurred since the last request.

In combination, this leads to lengthy retrieval times for RMON data. The reality is that the bandwidth cost of retrieving the data is so high that the majority of RMON users are unlikely ever to have an up-to-date view of the network, preferring instead to gather information after a problem has been notified. While RMON is a vendor-supported standard, it is bandwidth-hungry and processor-intensive, which inevitably compromises its usage and usefulness.

General RMON Architecture.   RMON is a widely adopted industry standard for the retrieval of network statistics from remote devices. RMON comprises two elements: an agent which is a remote "probe" attached to each segment to be monitored; and a client, which provides the management interface. The agent builds up information within 10 RMON groups. (See Fig. 6-3.)

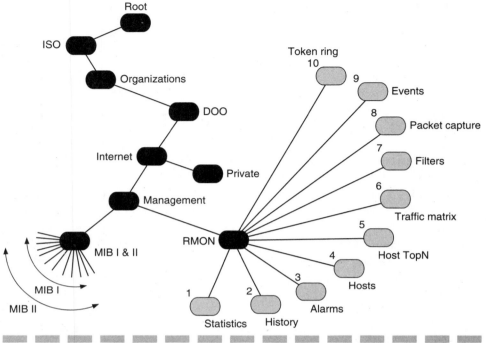

Figure 6-3 RMON standards and groups

While SNMP was designed around a passive polling architecture, RMON enables the use of "smart" agents which respond on an exceptions-only basis. This reduces the traffic associated with network management while still enabling the remote equipment to alert the SNMP management station when problems occur. By having the manager set thresholds, RMON allows probes to measure network performance. When the threshold for acceptable network behavior is exceeded, the RMON probe alerts the SNMP management station of the problem. The RMON protocol reports statistics at OSI layer 2 (the data-link layer).

Nondisruptive RMON probes dispersed across LAN segments work autonomously and report whenever an exceptional network event occurs. Filtering capabilities of the probe allow it to capture specific types of data based on user-defined parameters. When a probe discovers a segment in an abnormal state, it proactively contacts the RMON client application running on the centralized network management console and forwards to it the captured information describing the condition. The client application structures and analyzes the RMON data to diagnose the problem.

The RMON1 Architecture.   The RMON1 MIB was published in November 1991 by the Internet Engineering Task Force (IETF) to address the limitations of SNMP in growing distributed networks. The purpose of the RMON1 MIB is to allow SNMP to monitor remote devices both more efficiently and proactively.

The RMON1 MIB consists of a set of statistical, analytic, and diagnostic data which can be displayed using standard tools across multivendor product lines, thereby providing vendor-independent remote network analysis. The combination of RMON1 probes and RMON1 client software implements RMON1 in the network environment. The key to RMON1's monitoring effectiveness is its ability to store the history of statistical data at the probe, thereby removing the need for continual polling to build a view of network performance trends. Currently, the RMON1 standard only supports Ethernet and token-ring networks, but work is in progress on defining RMON standards for FDDI.

By allowing full support of all nine RMON1 groups (the tenth is specific to token ring), an RMON1 probe is able to establish event-driven communication to the network manager. This strategy allows for the off-line operation of the RMON1 probe from the SNMP management station either to minimize WAN communication links or to make sure that information is still connected when the SNMP manager cannot connect to the remote network—for example, during a network failure. The RMON1 probe provides for proactive monitoring of a remote network such that alarms can be signaled to the SNMP manager when events occur. The SNMP manager can then play back the event through the history logs on the RMON1 probe. One strength of RMON is that the probes can continually check for network problems and log them as they occur.

The RMON1 probe is directly connected to the LAN segment being monitored. In contrast, the SNMP manager is usually located on another segment. This allows the RMON1 probe to return a higher quality of information than the SNMP manager has access to. For example, the RMON1 probe can identify the top five nodes which are generating traffic. As these probes are located on distant LANs, it also gives them the capability to report statistics to several SNMP management stations within an organization at the same time.

RMON1 MIB Functional Groups.   The information RMON—retrospectively now called *RMON1* to distinguish it from the recently ratified RMON2 standard—provides for more comprehensive network management than would be available through other SNMP MIBs. The standard RMON1 MIB functions fall into 10 optional groups under three headings:

- Application-level activities
    - —*Filter*—packet-selection mechanism, matching packets to criteria
    - —*Packet capture*—packet collection and uploading mechanism through filtering criteria
    - —*Events*—control mechanism for actions triggered by an alarm
- Host statistics
    - —*Host*—host discovery and statistics by MAC address
    - —*Host TopN*—sorted, ranked statistic by MAC address
    - —*Matrix*—conversation tracking between two hosts
- Other groups
    - —*Statistics*—cumulative LAN traffic and error statistics
    - —*History*—interval sampling statistics for trend analysis
    - —*Alarms*—thresholding
    - —*Token ring*—four parameters for token-ring devices: ring station, ring station order, ring station configuration, and source routing statistics

Vendors which support any single RMON group or a subset of RMON groups are considered RMON-compliant. Filter and packet capture are the most processor-intensive and thus the most expensive to implement on switches (and thus often omitted). The two groups are also the most difficult to monitor on faster networks.

RMON1 in Detail. The RMON standard breaks the task of network monitoring down into nine basic groups, each providing a different type of information about the network.

Statistics group. Gathers information about the network's overall performance by measuring packets, bytes, broadcasts, multicasts, collisions, errors, and packet size distribution. This information forms the raw data which is used by the other groups' analytic functions.

History group. This group is used to specify sampling functions which define how often data is to be collected. The group creates a historical record of network performance by taking periodic samples of the Statistics group on the basis of user-defined sample functions and intervals.

Alarm group. Network managers use this group to predefine levels, or thresholds, for an event before an alarm is raised. They also specify the conditions under which an alarm will be raised. Alarm conditions can be based on changes (delta) or absolute value readings from any type of statistic.

Host group. This group gives network managers detailed monitoring information about hosts on the network. The group tracks statistics for each host on the network, sorted by its MAC address.

Host TopN group. This group extends the Host group information by sorting the information into lists. These lists can be sorted by statistics for example, the top hosts by packets sent, the top hosts by error packets, and so on. TopN can be used to place switches optimally in order to exploit backbone speeds.

Matrix group. The Matrix group gives the network manager specific information about traffic between pairs of nodes on the network. The output can be sorted by either source or destination MAC address.

Filter group. Using the Filter group, the network manager can tell the RMON monitor to zoom-in specific types of packets on an Ethernet segment. The group consists of a set of criteria under which packets will be forwarded to the Capture group and/or conditions which will trigger a log entry in the Events group.

Capture group. This group is used to set up a buffering scheme for capturing statistics which are sent from the Filter group. Buffers for holding packets are established, and captured groups will be forwarded to the network management station for analysis. The Capture group also allows for partial packets to be captured.

Events group. The network manager uses the Events group to define what type of network events are to be monitored. An event is triggered when RMON detects a predefined condition (above threshold) somewhere on the network. When detected, events can trigger actions and always generate a log entry for errors and user-defined conditions.

The 10th group was added later. The token-ring RMON specification, defined in the IETF Request for Comment 1513 (RFC1513), contains three groups specific to token-ring networks:

Statistics—tracks key events such as packets, bytes, ring purges, beacons, and nearest active upstream neighbor changes.

History—tracks utilization and errors over time. Typical reporting intervals are 30 s and 30 min.

Token ring—defines four subgroups specifically for monitoring token-ring networks. The *Ring Station* subgroup monitors events associated with each station on the local ring, and on the ring being monitored. The *Ring Station Order* subgroup tracks the order of stations on a given ring. The *Ring Station Configuration* subgroup allows active stations to be removed from the ring, and to download configuration information. The *Source Routing* subgroup tracks utilization on source route bridged networks, following statistics such as hop counts and broadcast frames.

In all other respects, the token-ring RMON specification is identical to its Ethernet counterpart, RFC1757. Both RFCs define groups for Alarms, Hosts, Host TopN, Matrix, Filters, Capture, and Events logging.

The Emerging RMON2 Standard.   The recently ratified RMON2 is an industry standard which enables network managers to monitor the network up to the application level of the network protocol stack. Thus, in addition to monitoring network traffic and capacity, RMON2 provides information about the amount of network bandwidth used by individual applications, a critical factor when troubleshooting client/server network environments. While RMON1 looks for physical problems on the network, RMON2 takes a higher-level view. It monitors actual network usage patterns.

RMON2 was developed to overcome the limitations of RMON1, which does not concern itself with the higher layers of the OSI model. Thus, RMON1 gives

- No indication of application protocols in use (e.g., IP, IPX)
- No statistics by application protocol
- No information on end-to-end host conversations
- No statistics by application

Whereas an RMON1 probe sees packets flowing from one router to another, RMON2 looks inside to see which server sent the packet, which user is destined to receive it, and what application that packet represents. Network managers can use this information to segregate users by their application bandwidth and response-time requirements, in much the same way as they have created workgroups by network address in the past. This capability helps greatly in defining appropriate VLANs by traffic flows.

TABLE 6-2

Network Management Focus of RMON1 and RMON2

Network management issue	Relevant OSI layer	Management standard
Physical errors and utilization	Media access control (MAC)	RMON1
LAN segmentation	Data link	RMON1
Interconnection of networks	Network	RMON2
Application usage	Application	RMON2

RMON2 is not a replacement for RMON1 but a complementary technology. RMON2 provides a new level of diagnosis and monitoring which builds on the RMON standard. In fact, RMON2 can use alarms and events from devices implementing standard RMON1.

In client/server networks, well-placed RMON2 probes can see the application conversations for the entire network. Good locations for RMON2 probes in a switched environment are in data center, backbone, or workgroup switches or in high-performance or key servers in the server farm. The reason is simple—that is where the majority of the application traffic passes. Physical problems are most likely to occur at the workgroup level, where users actually plug into the network. Thus, the workgroup is where current RMON1 implementations are most useful and most cost-effective, while RMON2 is better placed in the backbone of a switched network and the server farm. Table 6-2 summarizes the focus of each standard.

RMON2 Overview. RMON2 has been developed by the IETF to provide additional information for the troubleshooting at all communication layers. RMON3 is now under development—it will extend RMON to the WAN. RMON2 provides information required on an enterprise network giving end-to-end traffic statistics based on the major network layer protocols, such as IP, IPX, DECnet, AppleTalk, VINES, and OSI. Additionally, RMON2 provides application layer details, allowing network managers to troubleshoot and plan capacity based on applications such as Notes, Telnet, email, and databases.

RMON2 is a major enhancement to RMON1, and the RMON1 groups such as History, Alarms, Host, Matrix, and Filter are extended all the way to the application layer. Also provided are a User History group, address mapping of network layer to MAC layer, protocol directory and distribution, and a Configuration group. (See Fig. 6-4.)

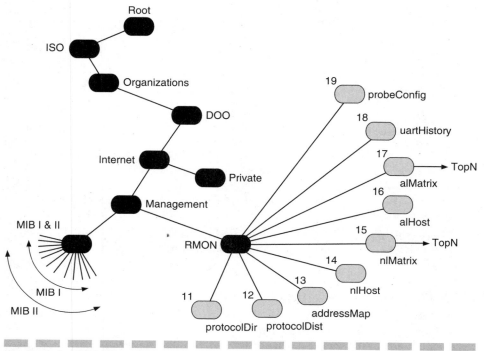

Figure 6-4   RMON2 groups

Key Goals.   The key goals of both RMON1 and RMON2 are to aid in the management of distributed LANs from a central site and to

- Provide a framework for proactive monitoring
- Aid problem detection and reporting
- Add value-added data
- Allow multiple managers access to the data

Additional Groups.   The major additions to RMON2 are these groups:

- Protocol Directory
- Protocol Distribution
- Address Mapping
- Network Layer Host

- Network Layer Matrix
- Application Layer Host
- Application Layer Matrix
- User History
- Probe Configuration
- RMON Conformance

RMON1 Enhancements.   RMON2 also specifies some enhancements to RMON1. These enhancements are implemented by any RMON2 probe which also implements RMON1, but they do not add any requirement to existing RMON1 probes. The additions to RMON1 are

- Adding dropped frames and last-create-time conventions to each group.
- Changes to the filter group to allow filtering based on an offset of a particular protocol, even if the protocol headers are of a particular length.
- Filter and capture status bits will add bits for WAN media and generic media. The added bits are
  —Bit 6: used to identify the direction of a packet on WAN media (local or remote).
  —Bit 7: used on any media to flag any physical layer error.
  —Bit 8: used on any media to flag any short packet on the media.
  —Bit 9: used on any media to flag any long packet on the media.

Application Layer Definition.   In the RMON2 definitions, the term *application layer* does not specifically refer to the OSI layer 7. The application layer in RMON2 is used to identify any protocol above the network layer and includes transport layers such as TCP.

Group Descriptions

Protocol Directory.   Every RMON2 implementation will have the capability to interpret certain types of packets and identify their protocol type at multiple levels. The protocol directory presents an inventory of those protocol types that the probe is capable of monitoring, and allows the addition, deletion, and configuration of protocol types which are in this list. One concept which requires explanation is the "limited extensibility" of the protocol directory table. With the

RMON2 model protocols are detected by static software which has been written at implementation time. Therefore, as a matter of configuration, an implementation does not have the ability to suddenly learn how to parse new packet types. However, an implementation may be written in such a way that the software knows where the demultiplexing field is for a particular protocol, and that the decoding of the next layer up is table-driven. This works when the code has been written to accommodate it and can be extended no more than one level higher. This extensibility is called "limited extensibility" to emphasize these limitations. Limited extensibility can be useful; for example, suppose that an implementation understands how to decode IP packets on any of several Ethernet encapsulations and also how to interpret the IP protocol field to recognize UDP packets. The implementation may also know how to decode the UDP port number fields and may be table-driven so that among the many different UDP port numbers possible, it is configured to recognize port 161 as SNMP, port 53 as DNS, and port 69 as TFTP (Trivial File Transfer Protocol). The limited extensibility of the protocol directory table would allow an SNMP operation to create a table entry which would define an additional mapping for UDP which would then recognize UDP port 123 as NTP; the probe could then begin counting such packets. This limited extensibility is an option which an implementation can choose to allow or disallow for any protocol which has child protocols.

Protocol Distribution. This counts the number of packets and octets for the different protocols detected on the network segment as defined in the Protocol Directory Group.

Address Mapping. This lists the MAC addresses to the network addresses discovered by a probe for all the protocols that the probe has the ability to match as defined in the Protocol Directory Group.

Network Layer Host. This counts the amount of traffic sent to and from each network address discovered by the probe for all the protocols which the probe has the ability to match as defined in the protocol directory group. For each host, the number of packets and octets in and out to each host is recorded as is the total of multicast and broadcast packets transmitted by each station.

Network Layer Matrix. This counts the amount of traffic between each pair of network addresses discovered by the probe for all the protocols which the probe is able to match as defined in the Protocol Directory group. *Source destination* and *destination sources* tables are stored together with the number of packets and octets for each conversation discovered.

Matrix TopN. The Network Layer group also allows the running of TopN conversations over a user-defined period based on either packet or octet counts.

Application Layer Host. This counts the amount of traffic, by application protocol, sent to and from each network address discovered by the probe, for all protocols which the probe has the ability to match as defined in the Protocol Directory group. For each host, the number of packets and octets in and out to each host is recorded.

Application Layer Matrix. This counts the amount of traffic, by application protocol, between each pair of network addresses discovered by the probe for all the protocols which the probe has the ability to match as defined in the protocol directory group. Source destination and destination sources tables are stored together with the number of packets and octets for each conversation discovered.

Matrix TopN. The Application Layer group also allows the running of a TopN conversation over a user-defined period, by application protocol, based on either packet or octet counts.

User History. The User History group is similar to the RMON1 history group except that the variable(s) recorded are defined by the user. This allows the user to create custom history tables (e.g., by protocol type, MAC or network address). The user defines both the sampling time interval and whether it is an absolute value or a delta value (respectively, the value of the counter at the end of each interval or the change in value during each interval).

Probe Configuration. This group controls the configuration of various operating parameters of the probe in a standard manner. It includes information about the RMON groups which the probe supports; the hardware and software revisions of the probe; how to update the firmware; serial line configuration for out-of-band management, including modem setup commands, and configuration of each interface (IP address, community strings, trap destinations, etc.).

RMON Conformance. This group allows a manager to find out which groups and instances of a group a probe supports.

Switched Networks Need a New Perspective. However, the arrival of switches and VLANs changes the whole perspective for RMON. First, the Host group becomes irrelevant when there is only workstation (host) per port—where every segment is a "private" segment. Second, the Matrix group must be extended not only to provide a matrix within a segment but now also within the switch (port-to-port conversations) and between switches (interswitch conversations) in a matrix format. Similarly, TopN hosts—to be meaningful—must be extended to reveal TopN ports rather than end stations in a switched network.

RMON Placement. The workgroup and the backbone are the most likely network locations for high-performance switches. The criteria for deploying RMONs depend on the type of switched environment.

At the workgroup level, the management criteria for switches are similar to those for periphery hubs. Network managers need the ability to monitor the error rates of end-station nodes connected to dedicated switch ports to achieve maximum performance. For this area, switches and switching hubs usually support the four main RMON monitoring groups: Statistics, History, Alarms, and Events. Thus, RMON1 is sufficient for workgroup segments and switch monitoring. It is also reasonable to use RMON1 to monitor critical servers in the switched network.

In the backbone, physical layer monitoring becomes less important, because errors are locally isolated by the switch port and are not propagated to the backbone switch. So how can RMON be most effective in a switched backbone environment? To get a fuller picture of traffic flows in a switched network, individual high-speed RMON probes are attached to backbone switch ports to gain a comprehensive picture of network performance. However, this solution can be costly. When they become available, integral RMON2 probes in backbone switches can provide useful information for data flow, overall utilization, and trend analysis.

RMON in a Switched Environment

In a LAN-based environment, RMON helps network managers determine how to best segment their networks by keeping track of who is talking to whom. Through exception reporting, a network manager can identify heavy bandwidth users. These users can then be placed on their own network segment to minimize their impact on other users.

RMON agents can be implemented as standalone, dedicated probes, each of which monitors a single LAN segment. (See Fig. 6-5 for example of a probe on an analysis port.) Dedicated RMON probes can cost U.S.

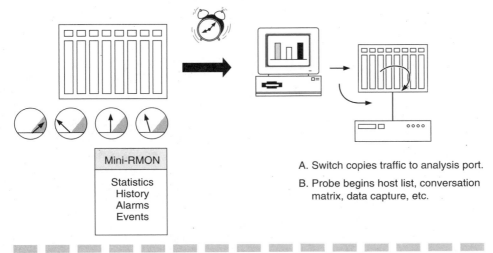

Mini-RMON

Statistics
History
Alarms
Events

A. Switch copies traffic to analysis port.

B. Probe begins host list, conversation
matrix, data capture, etc.

Figure 6-5   Probe on analysis port (Source: NetScout)

$3000 or more. However, while monitoring the network one segment at a time is of some use on that segment, it rather defeats the purpose of RMON. Proactive management of a network with RMON requires all segments to be monitored all the time. Depending on the size of the network, the cost of employing RMON probes across all LAN segments can be substantial. An alternative to using dedicated probes is to move probes physically from segment to segment to gain a systemwide perspective on network traffic patterns. However, this process is both difficult and time-consuming.

Multiport RMON probes provide information management capabilities on multiple LAN segments at the same time. They also provide the network manager a significant cost savings by eliminating the need to buy a single RMON probe for every segment. Before Ethernet switching became available, LANs were put together using hubs. All ports on a hub are connected to the same LAN segment, which means that all ports have access to all traffic. This means that all traffic on that segment can be monitored from any single port on the hub. This is good for monitoring, but results in a slower network because the bandwidth must also be shared within the segment.

Unfortunately, the bigger and more powerful a switched network becomes, the more difficult it gets to monitor and manage. Soon enough, the network could suffer from all the same old problems, bottlenecks and network congestion that switching technology was supposed to eliminate. However, with a switched network, problems could be harder than ever to locate and fix.

Switch Management

One disadvantage of a switched LAN environment is that it is hard to "see" inside the switches which make up the network. It is difficult to get the same level of management which is the norm for shared media LANs. This makes troubleshooting on switched networks more difficult.

Switch management determines how much control the network manager has over the switch. This can range from simple SNMP MIB I/II statistics, to viewing active ports on a hub, to altering the address tables and forwarding information.

Allegedly, one of the biggest reasons for implementing virtual LANs is to contain broadcasts and reduce the stress on the network. This is all very well, but how do you know that the design of the switched network and VLAN assignments has done just that?

Analyzing VLAN traffic is not easy, since the main reason for implementing VLANs is to create workgroups which are independent of the underlying physical network. For this reason, vendors are turning to RMON to give an indication of traffic flows. Most switch vendors implement the first RMON1 groups (Statistics, Alarms, Events, History) on each switch port. They also provide port mirroring (port tapping) to an external RMON probe or internal MIB or roving port/probe (roving RMON) which supports all nine RMON groups.

While both workgroup and backbone switches can be managed from an SNMP management platform, the workgroup switch is likely to support RMON1 while the backbone switch currently also supports RMON1 will eventually migrate to RMON2, which provides wider visibility of the network in real time.

External RMON Monitoring

In a switched environment, network managers now have a larger number of segments to worry about, and few Ethernet switches provide RMON capability. This creates a problem of how to manage many different network areas without having to buy an RMON probe for every new LAN segment.

Many switch vendors offer a "monitoring" port to allow a network analysis device to be connected to the switch, although their implementation techniques vary from vendor to vendor. There are three common implementation techniques which ease the introduction of external analysis devices:

- *Port tap*—sending all traffic to or from one port on a switch to a designated monitoring port on the hub
- *Circuit tap*—sending all traffic exchanged between two ports on the switch to a designated monitoring port on the hub
- *Switch tap*—sending all traffic which occurs on any port on the switch to a designated monitoring port on the switch

Each of these techniques simplifies network analysis and allows monitoring of a switch or connected device without the need for additional hardware. Many switches implement one or more of these techniques to assist analysis.

The port tap (often called *port mirroring* by Cabletron, SMC, Network Peripherals, and others) and circuit tap techniques limit the amount of traffic which is delivered to the network analyzer. This could filter important information from the analysis device if not used correctly. For example, an administrator may use the circuit tap to focus on a problem connection between a user and a resource. The source of the problem may actually be another user accessing the shared resource. Focusing on the connection between the first user and the resource may not provide enough detail to identify the problem.

At the other end of the spectrum, the switch tap does not filter any information. This may work well in networks where overall use is low. However, if several users are demanding high amounts of bandwidths individually—one of the reasons for moving to a switched environment—their combined traffic may be greater than the switch can process through one monitoring port.

Network Analyzers.   Network protocol analyzers have also adjusted to switched networks, offering the capability to monitor multiple segments simultaneously. The network analyzer provides a good complement to the RMON probe by offering more active features for fixing the problems which are detected on the network. The RMON probe is a tool for network management (an extension of SNMP), and the protocol analyzer is a tool for in-depth problem analysis. An RMON probe, for instance, can report a network problem. The network manager can then use the protocol analyzer to collect and analyze data packets to determine the cause of the problem and fix it.

External RMON—Early Solutions. Unlike LAN networks connected with standard hubs, switched networks offer point-to-point connections, which provide each user with a private 10-Mbit/s line. This means that each port handles traffic switched to and from that port only, without being affected by data on the other ports. While this promises significant performance improvements, it makes monitoring a nightmare. Early switch users found it difficult to pinpoint and eliminate congestion within LAN switches.

In a switched LAN network, each switch can act as its own separate LAN segment, making monitoring difficult. To mitigate this major problem, switch vendors together with network analyzer vendors provided three approaches, all using dedicated monitor ports on switches for analysis: roving monitors, diagnostic ports, and steering data to a remote site for further analysis. An alternative, but less useful, solution is to rely on SNMP information.

Roving Monitors. Roving RMON capability allows the network analyzer to focus more detail on a problem LAN segment. Typically, the network analyzer in standard operating mode monitors all connected LAN segments with four basic RMON groups:

- Statistics group
- History group
- Events group
- Alarms group

Roving RMON enables other RMON groups to be switched to a port of interest as required. The administrator can point roving RMON to provide additional information on a particular segment of interest. Typically, the additional information groups available with roving RMON are

- SNMP MIB II monitor proprietary extensions
- RMON Host group
- RMON HostTopN group
- RMON Matrix group
- RMON Filter group
- RMON Capture group

If a threshold is exceeded on one of the ports, a trigger can cause roving RMON to automatically focus attention on the problem port and signal the SNMP management station with a trap that the condition occurred. Triggers for roving RMON can be set for packets per second, bytes per second, errors per second, and total errors. Threshold can also be set for information contained in the SNMP Statistics group, which include CRC (cyclic redundancy check) alignment errors, undersized packet errors, oversized packet errors, jabber errors, and collisions.

The roving RMON probe provides the network administrator with the capability to simultaneously monitor multiple ports to provide a cost-effective method of tracking the performance of remote LAN segments. With roving RMON, the network analyzer can provide extensive information about problematic LAN segments. As well as reducing monitoring costs, roving RMON minimizes the performance impact of RMON on switch performance and simplifies network operation.

Diagnostic Ports.   A minihub or splitter can send copies of all signals from a port to an analyzing station on an unoccupied switch port. Only one port can be monitored at a time, forcing network managers to collate statistics gathered at different times from many sources. It also sacrifices a working port to the analyzer. A dedicated monitoring port—a diagnostic port—is provided on the LAN switch, but only one port can be monitored at a time. However, some devices can copy traffic from all ports (rather than selectively) to the monitoring port, which causes bottlenecks on heavily loaded systems.

A more sophisticated solution is to use RMON agents at each switch port. This provides complete information about the network, but when added to switches with a CPU architecture, switch performance is affected. The alternative, adding an external RMON agent at each switch port, becomes extremely expensive for switches.

Steering.   A partial solution to the potential bottleneck is to steer traffic to the remote site. This means that traffic is selected for delivery to a remote site. Multiple streams and conversations are copied to the port while packets are filtered at the remote analysis site. At the remote site, the data (TopN, Matrix, Filter, Capture groups) is correlated and integrated with data from other switches in order to monitor a workgroup consisting of multiple switches.

SNMP Statistics.   SNMP statistics can be used to measure traffic. Although standard in switches, SNMP provides only limited traffic information and fills the network with polling requests, as discussed in the previous section.

Disadvantages of External Monitoring.   The combination of switches together with embedded RMON on every port seems to be the answer to improving network speed, and allowing for flexible configuration. However, some switches on the market today lack RMON or any other mechanism for monitoring data as it travels over the network, so problems continue. Generally products which do support RMON can monitor only one port at a time because they use a special port dedicated for analyzers or RMON probes. This is like trying to measure attendance at a football match by counting the people coming through only one gate. All ports need to be monitored simultaneously, and there may be just too much data on the network to be monitored from a single port. To monitor a switched network, RMON can be built into the switch's hardware, but to be useful, it must be able to monitor all ports simultaneously.

RMON is available as an add-on to many switching systems, but only a few switches include embedded hardware-based RMON with every switching unit. However, there are drawbacks to using add-on RMON, or other monitoring systems placed onto networks. Often some probes can slow down the network, provide incomplete monitoring, or both. However, high-speed RMON probes are available, but they are a costly investment.

Embedded RMON

For RMON to operate effectively, each segment must have its own RMON agent to gather data for that segment. This presents four main challenges for designing RMON to fit into switched networks:

- Minimize the cost of each agent
- Report all required data
- Prevent RMON data from overloading the switch or network
- Keep management flexible to take advantage of the new switching capabilities

Embedded RMON (see Fig. 6-6), together with an efficient switch design, provides the greatest performance boost at the lowest price. The best way to do this without reducing switch and network performance is to use switches with embedded hardware-based RMON which can monitor all ports simultaneously. Several vendors have included a dedicated application-specific integrated circuit (ASIC) in the switch to provide RMON monitoring. This allows monitoring on all ports without reduc-

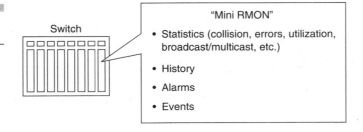

Figure 6-6
Embedded RMON

ing network performance or requiring any additional hardware investment. The intelligent RMON ASIC can be configured to gather data on any or all ports in the switch. Periodic reports in the form of summarized statistics and special warnings are sent to the network manager's monitoring station. These summarized reports make sure that the information is easy for the manager to understand, and prevent monitoring data from overloading the network.

The use of dedicated ASICs avoids placing the burden of RMON processing on the switch's CPU. In addition, ASICs can be added to a switch fairly cheaply in silicon. Some vendors have added ASICs on a per port basis, while others use the same ASIC to monitor several ports. The network manager can configure the RMON circuit to identify specific warning conditions. The manager will then be notified as soon as any of these conditions appear on the network, and will thus be able to fix the potential problem before network performance is affected.

Sometimes a particular port or set of ports is causing problems. By studying only those ports more closely, problems can be solved more quickly without gathering unnecessary information about other trouble-free ports.

However, embedded RMON has its disadvantages. Despite vendors' hype, embedded RMON implies full RMON support only on a per port basis, but you will wait in vain for the full set of RMON groups to be implemented in silicon on each port of a switch.

Embedded RMON takes the lowest-common-denominator approach to RMON monitoring. Typically, the ASICs monitor only four RMON groups on a port: Statistics, History, Alarms, and Events. For this reason, embedded RMON is sometimes scathingly referred to as "mini-RMON." This is purely because these groups are the easiest and cheapest to implement in silicon.

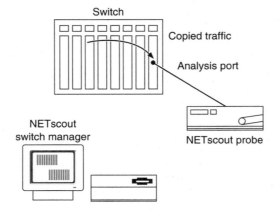

Figure 6-7
Roving probe
(Source: NetScout)

In addition, these groups have no performance implication for the switch—which would be the case if, for example, a vendor decided to implement the Packet Capture group on a per port basis. Vendors do not implement full RMON per port because of the impact it would have on switch performance and price/performance.

For this reason, embedded RMON needs the support of a full RMON probe—often referred to as "proxy" RMON—to be pointed at the port or internal circuit in order to gain more useful data about the size and scale of the potential problem flagged initially by the RMON subgroup data by copying data to a diagnostics port; hence, the term *roving RMON* or *roving probe.* (See Fig. 6-7.)

More importantly, switch vendors have yet to deal coherently with multiswitch workgroup information and correlate it to provide a higher level view of workgroup-backbone traffic based on RMON information. This requires an integrated traffic management approach to switched network and VLAN monitoring, correlating both internal and external workgroup switch traffic flows with application-related traffic flowing through the switched backbone of frame-based and ATM switches. The traffic management console must be able to aggregate traffic in meaningful ways in order to drill down to the source of congestion on the network.

Ideally, the traffic management system must be able to correlate and aggregate RMON data from multiple sources without causing congestion of the switched network. Further, in an ideal world, RMON data would be combined with ATM traffic management information (which is very different to RMON)—in a meaningful way—to reveal potential congestion in a mixed frame- and cell-switching environment.

For enterprise switched networks, RMON consoles should provide details—hopefully in a highly visible manner—of traffic aggregation at all levels of the switched VLAN network.

- By MAC address
- By subnet
- By network
- By ATM permanent virtual circuit (PVC)
- By switch
- By port
- By trunk
- By VLAN

Monitoring and managing virtual LANs (see Fig. 6-8) have hidden costs in traffic and network management. While RMON1 and RMON2 will enable traffic flows to be monitored and graphed by port or switch or VLAN, each workgroup switch must have embedded RMON on each

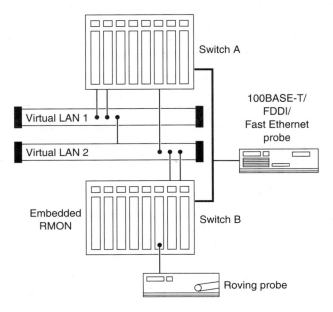

Figure 6-8
Managing virtual LANs

port together with an external roving RMON probe. In addition, in a multiswitch environment, a high-performance standalone RMON2 probe will be required on every high-speed backbone link where virtual LANs traverse backbone switches in order to track application traffic flows.

ATM Switching in Traffic Management

When virtual LAN traffic crosses ATM backbone switches, more information has to be gathered to make VLANs manageable. Traffic management is a fundamental aspect of ATM and how an ATM switch's call admissions and queuing algorithms affect its performance. Specific traffic-related information (i.e., virtual circuit statistics) of individual switches in the backbone is dependent on the design of the switches and which particular congestion algorithm has been implemented by the ATM switch vendor.

ATM handles these differing requirements by allowing different virtual circuits on one fast physical link to have different priorities and different bandwidth guarantees. Other technologies, such as Fast Ethernet, are just that—fast. All the traffic in a Fast Ethernet link has the same priority and contends equally for the available bandwidth. What is traffic management, and why is it needed? ATM *traffic management* is a framework for managing and controlling traffic and congestion in an ATM network to achieve the desired network performance and utilization objectives. In essence, traffic management, if carried out correctly, allows ATM networks to support traffic with varying performance requirements and fully utilize the available network bandwidth at the same time. The traffic management tools must be used carefully to be effective, especially for VLANs crossing an enterprise ATM backbone.

It is important to understand that if these tools are not used correctly, the ATM switch will not satisfy the performance requirements and/or the network utilization objectives for VLANs. For example, allocating quality of service (QoS) parameters to the virtual circuit(s) [VC(s)] which constitute part of the virtual LAN as well as the synchronization of VC forwarding tables have implication for the integrity and stability of VLANs as well as network performance. As mentioned before (see Chap. 4), the ATM Forum's LAN Emulation 1.0 is not considered a virtual

LAN, since this section focuses on the transport of layer 3 protocols constituting a VLAN across an ATM switch backbone, but an Emulated LAN (ELAN).

Information of particular importance for VLAN implementation across ATM switches include

- Forwarding table
- Cell counts per virtual channel (VC)
- Congestion per VC
- Congestion per port
- Buffer capacity

These basic statistics feed into the vendor's connection management system, which enables routes across the switch to be defined for the mapping of ports to virtual circuits as well as QoS metrics.

Forwarding Tables

The VLAN association tables (VC forwarding tables) must be synchronized across all switches. It is imperative for the layer 3 protocols running inside the VCs (the ELANs) that the broadcast domains not be breached under heavy traffic conditions (i.e., congestion) or by incorrect or outdated entries in the VLAN association table resident in each switch.

Whenever a device is mapped to several VLANs, it occupies space in multiple VC forwarding tables. For example, if 100 devices need to be mapped to four VLANs, then they will occupy 400 VC ID slots in the VC forwarding tables. If this is repeated for many other devices, then the VC forwarding tables will grow quickly.

The critical requirement for an individual backbone ATM switch is that it must be able to support a large number of VLANs. While it can be argued that edge switches serving individual wiring closets do not need to support as many VLANs as there are in the overall switched network, this cannot be applied to backbone switches which must transport VLANs from one side of the switched domain to the other, particularly if one of the main reasons for having a VLAN is to reduce the costs of moves, adds, and changes. This requirement becomes even more vital if policy-based VLANs have been implemented across the switched environment, since their sophistication means that even more VLANs have to be transported across the backbone.

Switches deployed in the backbone of the network often act as transit switches for VLANs which span the physical network but have no VLAN members mapped to the ports on the backbone switch. In such a scenario, the switch must be capable of handling high-volume traffic streams from multiple high speed (≥100-Mbit/s) trunk uplinks from edge or workgroup switches; and also retain VLAN integrity and stability.

If the switch becomes overloaded (congested), then cells or cell streams representing layer 3 VLAN packets will be missed or dropped, leading to the VLAN becoming unstable. When this happens, the non-ATM switched network is suffering "transit VLAN starvation," or, in the case of ATM networks, "VC starvation" (commonly called "beatdown"). A switched ATM backbone must be able to support large numbers of VLANs without performance (congestion) and stability impacts, in addition to any other traffic it may be carrying.

Congestion arises within an ATM network when the total rate of input of data into the network exceeds the available link capacity. There are several sources of congestion, including

- The bursty nature of the data traffic
- Unpredictable demands on the various network resources
- The heterogeneous nature of networks, with its attendant mismatch of link speeds
- The magnitude of propagation delays

VLANs Crossing Wide Area ATM Networks

Within an ATM WAN, VCs traversing large hops often receive poor performance as a result of propagation delays. This situation can be improved if a higher priority is given to existing traffic, rather than to data entering the network, via a network access control scheme.

ATM Traffic Management Basics

An ATM switch typically provides a traffic management facility which allows network administrators to maximize available network resources. The traffic management facility provides control over network resource allocation and ensures efficient use of resources that have not been explicitly allocated. This traffic management facility controls two key aspects of the QoS on every VC:

- Delay
- Bandwidth availability

Delay is controlled by a mechanism called *transmit priority.* Delay-sensitive traffic can be given preferential treatment in an ATM network if assigned a higher transmit priority. Bandwidth availability is controlled by four complementary mechanisms which operate at different levels in the network:

- *Bandwidth allocation*—keeps track of the amount of bandwidth that has been reserved for each VC.
- *Call admission control*—prevents network users from allocating more bandwidth than the network can provide.
- *Traffic policing*—operates at the edges of the network to ensure that, once a VC has been established, it does not try to use more bandwidth than the network currently has available.
- *Selective cell discard*—deals with momentary oversubscription of a trunk or edge port. When a traffic surge exceeds the buffer capacity at an output port, this mechanism selectively discards cells, giving preference to different classes of traffic according to parameters set by the network administrator.

The bandwidth availability mechanisms are supported by two additional traffic management features:

- *Rate-based congestion avoidance system*—keeps the traffic policers on edge modules informed about how much bandwidth is currently available in the network, so that they admit only traffic that has a high probability of being delivered
- *Traffic shaping*—meters incoming packet traffic to reduce the occurrence of surges that could exceed the buffer capacity on any output port

ATM Forum Traffic Management Tools

The main tools defined by the ATM Forum to manage traffic are the QoS parameters, the traffic service categories, the usage parameter control (UPC) function, and the call admission control (CAC) function.

Quality of Service Parameters. The ATM Forum has defined three negotiated QoS parameters: the cell transfer delay (CTD), the cell delay variation (CDV), and the cell loss ratio (CLR). Other parameters exist but are not negotiated and are not discussed here. CTD is defined as the maximum end-to-end cell transit time. The CTD is a function of the transmission delays and the ATM switch queuing delays. CDV is the amount of time between the maximum end-to-end cell transit time and the minimum end-to-end cell transit time. CDV is a function of multiplexing many connections onto a single physical link and the variability in ATM switching queuing delays.

Finally, the CLR is defined as the number of lost cells divided by the number of total transmitted cells. The CLR is a function of the physical link's error rate and the congestion management algorithms used in an ATM switch. As you can see, the switch congestion management and queuing algorithms play a critical role in the performance an ATM network can deliver.

Traffic Parameters. The traffic management specification defines several parameters used to describe the shape of the traffic. These traffic shapes in conjunction with the QoS parameters define the ATM layer service categories. The major traffic parameters are described in Table 6-3. Other parameters exist for the available bit rate service category. They are collectively described as feedback parameters and are not described here.

Feedback Parameters. These are a set of parameters used by the available bit rate service category to inform the source of the network resources available to it. The two main feedback mechanisms are explicit forward congestion indication (EFCI) and explicit rate (ER).

TABLE 6-3

Major Traffic Parameters

Acronym	Traffic parameter	Description
PCR	Peak cell rate	The maximum cell rate a source is allowed to maintain
SCR	Sustainable cell rate	The average cell rate a source is allowed to maintain
MBS	Maximum burst size	The number of cells a source is allowed to send at the PCR
MCR	Minimum cell rate	The minimum cell rate a source is allowed to maintain

Service Categories

The ATM Forum has defined five service categories: constant bit rate (CBR), real-time variable bit rate (rtVBR), non-real-time variable bit rate (nrtVBR), unspecified bit rate (UBR), and available bit rate (ABR). Table 6-4 describes how the service categories are related to the various types of traffic that could be transported over an ATM network. Table 6-5 shows the parameters that are specified for each of the defined service categories. Both voice and video require QoS parameters that are specified for CBR and rtVBR traffic.

Constant Bit Rate.   Since the rate of CBR traffic is very predictable, CBR traffic is the easiest type of traffic for which QoS parameters can be guaranteed. There is no feedback for this type of service.

Variable Bit Rate.   VBR traffic is characterized by having PCR (see acronyms listed in Table 6-3) at which you can send the MBS number of cells, followed by an SCR, which can be maintained indefinitely. VBR traffic is defined in such a way that the average rate of the VBR traffic is always less than or equal to the SCR. There is no feedback for this type of service. The only difference between real-time VBR (rtVBR) and non-real-time VBR (nrtVBR) is whether the QoS parameters are specified.

Unspecified Bit Rate.   UBR has no QoS or traffic shaping guarantees at the ATM level. To use this service, a higher-layer protocol such as TCP must detect and react to congestion. For instance, TCP can adjust the transmission rate according to the number of packets lost in transit.

Available Bit Rate.   One set of strategies for dealing with congestion is referred to as *available bit rate* (ABR) flow control. ABR strategies involve intermediate switches and/or final destinations to signal sources of data to slow down transmission when traffic exceeds capacity. In addition to providing effective and efficient responses to the causes of congestion, ABR flow control schemes provide dynamic allocation of available bandwidth and address differing CLR requirements.

The ABR service category has been the main focus of the ATM Forum's Traffic Management 4.0 document. The service category's basic assumption is to provide feedback directly to the source of the ABR traffic indicating how much network resources are currently available. Because ATM links operate at reasonably high speeds, the ABR feedback-response mechanism must be implemented in hardware. An ATM switch can provide three types of feedback: explicit forward congestion indication (EFCI), explicit rate (ER), and virtual source virtual destination (VS/VD).

The ABR service category requires that a resource management (RM) cell be inserted in the traffic flow periodically (usually twice every 32 data cells). This RM cell is used to convey congestion information back to the ABR source. Cells heading in the direction of the data are called *forward resource management* (FRM) cells, and cells heading in the opposite direction of the data are called *backward resource management* (BRM) cells.

The ABR ER Feedback Mechanism

The VS/VD feedback mechanism is identical to the ER mechanism except that each VD can convert FRM cells into BRM cells and each VS must generate FRM cells and react to BRM cells. The benefit of adding this complexity into the network is that the feedback's round-trip time is reduced to the time between VS/VD pairs instead of being the end-to-end delay. Each VD must have a corresponding VS. The VD-to-VS communications is implementation-specific and has not been standardized. Another advantage of the VS/VD function is that it can obviate the need for an ABR policing function. Since the VS/VD pair adheres the ATM Forum's ABR source and destination protocols, a VS/VD pair also can be used to ensure ABR compliance.

Usage Parameter Control

The UPC function is used to check to see whether individual VCs or VPs are adhering to their traffic parameters. The UPC function eliminates the possibility of unauthorized use of available network bandwidth.

Call Admission Control

The CAC function checks to see whether network resources are available to support the QoS and traffic parameters of an incoming call. The main network resources are the outgoing line's transmission capacity and the switch's switching capacity. The exact definition of capacity depends on the service class and the values of QoS parameters being supported. There follows an example of how the CAC could be used to guarantee network performance for the service classes requiring specific QoS parameters.

How to Guarantee ATM Network Performance

As described previously, the ATM Forum has defined five service categories with varying performance requirements. One of the hardest jobs an ATM switch must perform is choosing which cell will be sent next. The easiest way to accomplish this selection is to have a strict priority scheme.

In this scheme, if any cells are available for CBR traffic, they will be transmitted first. If no cells are available for CBR, then rtVBR cells will be transmitted. If no cells are available for either CBR or rtVBR, then nrtVBR cells will be transmitted, and so on until all service categories are examined. The strict priority scheme suffers from a starvation problem. Since CBR traffic never produces a burst, having a strict priority is acceptable because there will always be bandwidth available for the other service categories. However, if VBR traffic is given a strict priority over other classes, then several VBR VCs may start transmitting at their PCRs. The sum of these PCRs could exceed the available bandwidth, and the service categories with lower priority might be unable to get any cells out.

This could cause timeouts in the higher-layer applications, such as a layer 3 VLAN, which would then have to retransmit the data contributing to even further congestion. The problems worsen if ABR and UBR are prioritized. Since the ABR and UBR service categories have been designed to utilize all the available bandwidth, the lower-priority service category will constantly be in starvation. Recovery time must be tested and defaults set.

Strict Priority Servicing Scheme

A more complex but less problematic way of selecting which cell will be sent next is to have strict priority for the CBR service category (to provide a very high QoS) and some sort of weighted minimum bandwidth allocation scheme for the other service categories. Ideally, the weighted minimum percentages should be adjusted as calls are set up and torn down to reflect the desired bandwidths of the VCs in each of the service categories.

Queuing Discipline

An ATM switch's queuing discipline directly determines how effectively it will utilize the excess network bandwidth. The various queuing algorithms are discussed in Table 6-4. Typically, first-generation ATM switches only supported per VC accounting while second-generation ATM switches supported both per VC accounting and per VC queuing.

Since UBR has no QoS or traffic shaping guarantees at the ATM level, to use this service, a higher-layer protocol such as TCP must detect and react to congestion. TCP can, for example, adjust the transmission according to the number of packets lost in transit. Since the TCP protocol can react only to whole packets and since dropping one cell in the packet causes a retransmission of the packet, UBR without early packet discard (EPD) produces very disappointing results. In early packet discard, if a cell needs to be dropped, the whole packet is dropped except for the very last cell. The last cell in the packet is not dropped in order to allow TCP to detect a packet error and request a transmission immediately.

Table 6-5 describes the various queuing algorithms and their effectiveness for UBR with TCP. A "Very low" in the "Excess bandwidth utilization" column represents 0 to 20 percent of the excess bandwidth. A "High" represents 70 to 90 percent of the excess bandwidth.

The queuing algorithm used by the switching system directly affects ABR's ability to use the excess bandwidth. The efficiency of the various

TABLE 6-4

Switch Queuing Algorithms

Queuing algorithm	Description
FIFO queuing	All VCs in the same service category are placed in the same queue; all VCs are equally affected by congestion in the service category
Per VC accounting with FIFO* queuing	All VCs in the same service category are placed in the same queue, congestion is detected on a per VC basis—congestion on one VC can still affect other VCs because they share the same queue
Per VC queuing	Each VC has its own queue; congestion on one VC does not affect other VCs
UBR with TCP	UBR has no QoS or traffic shaping guarantees at the ATM level

*First-in, first-out.

TABLE 6-5

Effectiveness of
Switch Queuing
Algorithms for UBR
with TCP

Queuing algorithm	Advantages and disadvantages	Excess bandwidth utilization
FIFO queuing with no EPD	Many useless cells in the network and many retransmissions cause congestion collapse; many VCs will experience 0% throughput because of beatdown (alternating congestion on switches along the path)	Very low
FIFO queuing with EPD	No fairness among the VCs; many VCs will experience 0% throughput because of beatdown	Low
Per VC accounting with EPD	Fairness among the VCs; congestion on one VC can still affect other VCs	Medium
Per VC queuing with EPD	Fairness among the VCs; congestion is limited to each VC	High

queuing algorithms and the feedback mechanisms is displayed in Table 6-6. A "Very low" in the "Excess bandwidth utilization" column represents 0 to 20 percent of the excess bandwidth. A "Very high" represents greater than 95 percent of the excess bandwidth. The table clearly shows that as the feedback and queuing algorithms get more specific, the excess bandwidth utilization increases. It should be noted, however, that even when ER-based per VC feedback is applied, poor bandwidth utilization can occur if the algorithm used to calculate the desired ER is not a robust algorithm. A robust ER algorithm must perform buffer management, rate management, and fair-share bandwidth allocation.

The fact is that all EFCI control schemes propagate a bias against "long hop" VCs. A VC that passes through a relatively large number of links—which is highly likely for a VLAN straddling an ATM backbone—has a greater chance of running into congestion, which leads to more set EFCI bits, which in turn leads to fewer chances for the source to increase the associated allowed cell rate (ACR). This phenomenon of VC starvation is commonly known as the "beatdown problem."

To support the QoS guarantees required by voice and video information streams, a switch's connection admission control algorithm needs to allocate extra bandwidth to those channels. Both ABR and UBR with TCP can be used to harvest the extra bandwidth which is not actually used by the voice and video channels. The ATM switch's queuing algorithms play a critical role in determining QoS of the channels and how

TABLE 6-6

Effectiveness of
Switch Queuing
Algorithms for ABR

Queuing algorithm	Advantages and disadvantages	Excess bandwidth utilization
FIFO queuing with EFCI	No fairness among the VCs; many VCs will experience 0% throughput due to beatdown; feedback is inefficient (binary) and depends on round-trip time	Very low
Per VC accounting with EFCI fairness among VCs	Congestion on one VC can still affect other VCs; feedback is inefficient (binary) and feedback depends on round-trip time	Low
Per VC queuing with EFCI fairness among VCs	Congestion is limited to each VC; feedback is efficient and depends on round-trip time	Medium
Per VC accounting with ER fairness among VCs	Congestion on one VC can still affect other VCs; feedback is efficient (explicit rate) and depends on round-trip time	Medium
Per VC queuing with ER fairness among VCs	Congestion is limited to each VC; feedback is efficient (explicit rate) and depends on round-trip time	High
Per VC queuing with VS/VD fairness among VCs	Congestion is limited to each VC; feedback is inefficient (explicit rate) and is reduced to the time between VS/VDs	Very high

much of the excess bandwidth is actually recovered. The per VC queuing algorithms provide the highest QoS and network utilization.

Virtual Sources and Destinations

The problem of potential large delays associated with long-hop VCs can be fixed by segmenting an ATM network into smaller pieces, each containing one switch. Within each segment, the switch would act as a virtual source and/or virtual destination relative to the other elements within the segment. This setup would create smaller feedback loops, and thus faster response time. Each segment would be allowed to use any available congestion control scheme. Also, "misbehaving" users would be isolated within the first control loop of the network. The disadvantage is that a switch in this arrangement, referred to as a *VS/VD* switch, needs to maintain a queue for each VC, which adds considerable expense. Multicast VCsExplicit rate feedback schemes can be extended to work for point-to-multipoint (P2MP) connections as well as for point-to-point (P2P) connections. In this arrangement, branchpoint devices replicate cells traveling from the root of the P2MP tree to the leaves and consolidate feedback information traveling in the reverse direction. As long as branch traffic conforms to the expected behavior for a P2P connection in each case, actual and virtual sources and destinations, and switches, would behave the same way for P2MP connections as they would for P2P connections.

Congestion Control Scheme for Wide Area ATM Networks

Within a WAN, virtual circuits traversing large hops often receive poor performance as a result of propagation delays. This can be improved if priority is given to existing traffic, relative to data entering the network, via a network access control scheme. The combination of access control and rate control enables VCs extending over all distances to receive comparable (excellent) performance interoperability of EFCI and ER switches. The beatdown problem tends to be associated with switches implementing EFCI control schemes. Experiments have demonstrated that each time an EFCI switch is replaced by an ER switch within a particular ATM network, performance—throughput and "fairness" of bandwidth allocation—improves to a measurable extent.

Is ATM Manageable?

One of ATM's major challenges is network management. The five-layer network management model created by the ATM Forum provides a structured approach for managing ATM networks.

The Network Management Working Group of the ATM Forum is finalizing this end-to-end management model, which will be applicable for both private and public networks. The model will also define gateways between SNMP management systems (e.g., Sun's Solstice SunNet Manager), CMIP (Common Management Information Protocol), and existing proprietary systems. It will also define special OAM (Operations, Administration, and Management) cells to distribute management information over ATM networks. Five key management interfaces, labeled M1 to M5, are defined in this framework:

- M1 and M2, which define the interface between the management system at the customer site and the ATM end station
- M3, which defines the CNM (customer network management) interface
- M4, which defines the merger of private and public networking technologies, enabling NML (network management level) and EML (element management level) views
- M5, the management interface between a carrier's own network management systems

Special status in this framework is accorded to SNMP—the recognized standard for LAN network management. M1 and M2 interfaces encompass the relevant standard MIBSs (management information bases) for DS-1, DS-3, SONET (Synchronous Optical NETwork), and the AToM MIB—as defined by RFC1695. The AToM MIB allows network managers to group collections of switches, virtual connections, interfaces, and services into discrete entities and is SNMPv1/v2-compliant. It is expected to reduce the forest of exiting proprietary vendor-specific ATM MIBs.

In 1993, the Internet Engineering Task Force (IETF) formed an AToM MIB Working Group to define and develop the necessary management objects using the SNMP protocol for managing ATM devices. The current version of the AToM MIB (otherwise known as IETF RFC1695, *Definitions of Managed Objects for ATM Management,* version 8.0) defines object groups that can help net management applications construct some, but not all, of these views of the network. It specifies several SNMP management objects to manage ATM interfaces, ATM virtual links, ATM crossconnects, AAL5 entities, and AAL5 connections supported by ATM hosts, ATM switches, and ATM networks. The MIB uses SNMP SMI (Structure of Management Information) version 2 to describe these objects, but it can be accessed by SNMPv1 management applications.

Although the AToM MIB provides much of the same information as the ILMI MIB, it is less function-rich. Currently, the AToM MIB is used primarily to manage ATM PVCs. It can handle partial management of SVCs, but full management—including signaling functions—will require capabilities beyond the scope of RFC1695.

Also, the AToM MIB does not define objects needed for ATM service management or for a true end-to-end view of the ATM network. One reason for this is that the MIB is limited to a single ATM end system or switch and does not support the whole end-to-end span of a VC (or VPC), which spans multiple ATM end systems and/or switches. Thus, the end-to-end management of a VC or VPC is achieved only by appropriate management of its individual segments in combination.

Since most LANs today are SNMP-manageable, and carriers are still using CMIP (as well as proprietary network management protocols, in many cases), the ATM Forum recognizes that private and public network management systems will maintain separate views into the network for some time. Therefore, the ATM Forum network management scheme will enable both views and involve the building of gateways between these dissimilar systems.

Other groups, such as the Network Management Forum and the ITU (International Telecommunications Union), have been working with the ATM Forum to negotiate acceptable interworking standards. Obviously, it will take time to finalize specifications for interworking between SNMP and CMIP. The good news is that some progress has been made with the M4 interface toward establishing a protocol-independent MIB, which will support SNMP objects based on SMI, as well as CMIP objects, conforming to GDMO (Guidelines for Development of Managed Objects). The bad news, however, is that there is a long way to go for full bidirectional interoperability between SNMP and CMIP.

In the next few years, development of ATM management is likely to become a priority. However, at least in the earlier stages, most developments are likely to be based on OAM cells, rather than on SNMP or CMIP architectures. The ATM Forum is currently specifying three types of 53-byte OAM cells with specialized identification and field tags, to identify their functions: fault management, performance management, and activation/deactivation (for starting and terminating fault and performance management functions). By providing ATM network devices with the ability to gather information about end-to-end connections, the OAM cells will reduce the need to distribute MIBs throughout the network, and significantly lighten the amount of management-related traffic.

Standards for Managing ATM Performance

Efficient network management is one of the major functions required in any networking architecture. What tools are available today for managing ATM networks?

RMON was developed for shared media (such as LANs) and does not scale well to switched architectures (including ATM). RMON, as well as protocol analyzers, needs an ATM probe to act as a passive observer, which can unobtrusively steer cells to where they can be reassembled into packets for debugging.

In July 1995, a group of manufacturers of ATM switches, test equipment, and RMON products announced formation of the AMON (ATM MONitoring) group to develop an AMON agent, which will route copies of troublesome virtual circuits to a test port, for monitoring ATM layers. While some ATM-specific MIBs already exist, the proposed ATM Circuit Steering MIB expands the contents higher in the ATM stack to include virtual paths, virtual circuits, destination ATM addresses, and timers.

AMON capabilities and objectives of the ATM consortium include

- A nonintrusive way of monitoring virtual circuits of interest
- An upgrade path for the existing monitoring equipment
- A standard method of steering virtual circuits to monitoring, analysis, and troubleshooting equipment
- Accelerating time to market and increasing the functionality of ATM devices and monitoring equipment

However, additional standards are needed to define what data should be collected and how to present that data to a network management system. An interesting development in this arena recently came from a major player, Cisco Systems. In February 1996, Cisco, with a group of RMON and analysis tools manufacturers, developed a new draft for ATM remote monitoring. This draft defines the means of by which an RMON agent, embedded on ATM switches and distributed across ATM networks, can gather network traffic information. It is based on the original RMON1 (RFC1271) and RMON2 (RFC1757) specifications.

As a twist, Cisco, one of ATM Forum's founders, submitted this draft to the IETF rather than the ATM Forum. But Cisco and six other vendors submitted it to the ATM Forum in late 1996. This ATM RMON draft defines different methods which can be used to collect network traffic information. One of the methods uses "circuit steering" as proposed by the AMON consortium. In addition to circuit steering, the draft defines the option of having cell-level RMON instrumentation integrated in an ATM switch fabric. To size ATM traffic usage, the draft employs basic cell statistics per port, per virtual connection, per host, and per conversation pair.

Such detailed information can help network managers determine the performance bottlenecks and find the top talkers in an ATM network. It

is quite challenging to capture the data at full rate because of ATM's 155-
to 622-Mbit/s network fabric and the limited amount of available storage
(RAM). Therefore, the specifications also discuss data sampling methods
which allow capture of frames, rather than random cells—which would
yield little useful information.

ATM Management Tools

Management of ATM networks is a challenge. Since most of the standards are still work-in-progress today, the best advice is to use proprietary tools specifically designed for their hardware, and address ATM
network management with carriers and public service providers as early
as possible during contract negotiation. It is also probably a good idea to
request a direct feed from the carrier's network management system
into your network management center.

Most proprietary tools for ATM network management will run on
the major SNMP network management platforms, such as Sun's Solstice
SunNet Manager. Tools include

- ATMDirector from Cisco Systems
- Effective Networks' ENsite Adapter Manager
- ForeView from FORE Systems
- Northern Telecom's Magellan Network Management System, which
  runs under Solaris or SunOS and can be integrated with Solstice
  SunNet Manager and others

There are also hardware-based tools and equipment that fall mainly into
two categories:

- High-end, ATM-only analyzers
- Generic analyzers, offering ATM as one of their features

The first category of high-end ATM analyzers usually comes with
advanced traffic generators that can capture and replicate ATM activity
correlating to several PVCs.

SNMP and ATM

SNMP was not designed to handle the huge volumes of network traffic
ATM generates. SNMP-based tools have been extended to support ATM

networks. To accomplish this, network management architecture for ATM switched networks is decentralized and deployed as a series of agents which report from various ATM devices when a connection is established. By using this approach, ATM devices can send SNMP data to a central network management station, thus providing SNMP monitoring in real time. It also extends the role of network hardware. ATM switches are now playing a critical role in network management strategy. One of the most important conditions in selecting an ATM switch is to make sure that it supports a multiagent SNMP architecture to allow reconfiguration without losing its manageability.

Modeling

Simulation is a useful approach for predicting the performance of networks. Traditionally, simulation has traditionally been the job of mathematical experts who understand statistical theory and simulation languages.

Over recent years, network modeling has hidden its mathematical roots; and modern simulation toolsets make modeling simpler than in the past through the use of libraries of device parameters which represent specific network vendor devices. The accuracy of simulation results has also improved through the use of real data from RMON probes—rather than crude assumptions about traffic behaviors. In addition, some network simulation tools build up the topology of the existing network through SNMP autodiscovery.

Network modeling helps to answer such questions as

- What happens if I add 200 more Lotus Notes users?
- What impact will consolidating servers have on network performance?
- What will be the impact of moving this workgroup to a new location?

What-if simulation capability allows evaluation of proposed network designs and changes before they are actually implemented. Starting with a model of an existing network, desired changes to the network can be analyzed to evaluate their impact and effect. LAN segments, workstations, servers, bridges, routers, switches, WAN links, an applications can be added, moved, deleted, and reconfigured. Traffic intensity can be increased and decreased. Simulations of new client/server applications such as VLANs can be designed to evaluate their impact on existing network traffic.

Intelligent optimization algorithms are used to present specific problem-solving recommendations which address network operation. On the basis of optimization criteria, network modeling software makes recommendations on

- Optimal server, switch, and workstation placement
- Capacity requirements for WAN links
- Interface speed requirements for servers: LAN segmentation, hub partitioning, LAN switch, and VLAN deployment
- Survivability (of bridges, routers, switches, and WAN links)

However, certain issues need to be understood if network simulation is to be used in the right circumstances and if the common disadvantages of simulation are to be avoided, especially in switched networks. What is *simulation?* Network simulation is a technique used to predict the performance of a network. Simulation uses a model of a network to predict its performance. To put simulation in perspective, consider the alternatives used to predict the performance of a network:

- Experiment with actual equipment
- Make changes to production network: analytic network modeling
- Discrete-event simulation

Approaches to Performance Prediction

Network performance could be predicted by constructing a test network in a laboratory and seeing how it performs. This approach is useful when investigating new network technologies (such as VLANs) or applications on a small scale. Its disadvantage is that it is very costly, in terms of both time and equipment, to scale such a laboratory test to the size of a real production network. However, laboratory testing is a valuable source for collecting detailed data which will be used in simulation.

A second approach is to make changes to the production network. This is a very risky approach because of its potential to bring down the network and cost an organization millions of dollars in lost productivity. Some changes to a production network are less risky—such as a pilot project of a new application. Properly handled, a pilot project will not interfere with normal network operations yet will have the benefit of testing the application and network interactions in an environment similar to the production environment. Both laboratory tests and pilot projects can be valuable sources of data for network modeling.

Network Modeling

Analytic.   Two other approaches for predicting network performance both involve modeling the network. The first approach, *analytic modeling,* involves creating a number of equations which describe the network's performance and solving them for the unknowns such as utilization and latency.

Analytic modeling has the benefit that it is computationally very fast and can be used to solve even a large model. Its main disadvantage is that it is not able to account for all the complex behaviors which applications and network devices exhibit. The ability of analytic modeling to accurately predict latencies is limited because of the simplifying assumptions often made to make the mathematical analysis amenable to solution.

Nevertheless, analytic modeling is very useful for network capacity planning because

■ Many network capacity planning decisions can be made without a high degree of precision. For example, if an analytic approximation concludes that a WAN bandwidth of 87.3 kbits/s is required, the link would be sized at the next-highest capacity—a fractional T1 at 128 kbits/s.

■ The network traffic (load) measured at a given time is a snapshot: variations are due to the inherent randomness in the network traffic—network traffic tends to grow over time. Consequently, most networks are designed with some headroom for traffic growth. This means that approximations will still yield useful results.

Discrete-Event Simulation.   Discrete-event simulation uses a model where the traffic is represented as sequences of messages, packets, or frames. The movement of the traffic through the network is simulated by keeping track of the state of the network devices and traffic sources as they evolve over time.

Discrete-event simulation uses a central clock which moves forward in time at distinct points during the simulation, such as when a packet is generated by a node or transmitted by a router. Discrete-event simulation requires a time profile of the creation of traffic.

Traffic is often simulated using random arrival times because of the inherent randomness of the traffic. The models of devices (bridges, routers, switches, servers) simulate the queuing and processing of packets.

An important aspect of discrete-event simulation is the abstraction used both in the device and in traffic modeling. Every simulation model has a range of operation under which it is valid. A broader range of operation generally requires a more complex model. This results in more time to construct and execute the model. Therefore, it is important to focus on a certain subset of problems. The device models in network modeling simulate the factors in the devices that most affect network performance.

Simulation

Types of Simulation. Thus, there are two types of simulation: fast simulation and full simulation. *Fast simulation* uses a form of analytic modeling tailored to network capacity planning. Fast simulation calculates the average load placed on segments, devices, and WAN links based on conversation (TopN) information that is imported from an external data collector—external to the simulator, that is—such as RMON probes. The conversation information used in fast simulation is the rate (in bytes per second) of data transfer between a pair of nodes. A conversation is bidirectional, meaning that the rates in the two directions are not required to be equal.

Full simulation executes an event-driven simulation of the network model over a specified time period. The model is driven by traffic: sequences of client-to-server requests and server-to-client responses. Traffic is derived automatically from imported traffic conversations from RMON probes, and additional client/server traffic can be specified through a graphical interface.

The imported conversation information is converted into a random traffic pattern which will result in the same flows between two nodes if the network between them is not a bottleneck. When adding additional client/server traffic, this is specified as a transaction—the amount of data going from the client to the server and the amount of data returning—and the rate at which the client application sends these transactions to the server. The simplicity of fast simulation's mathematical analysis makes it best suited for network capacity planning on large net-

TABLE 6-7

Fast versus Full
Simulation

	Fast simulation	Full simulation
Modeling technology	Analytic (mathematical)	Discrete-event simulation

Primary uses	Capacity planning
	  Determining traffic flows
	  Determining traffic sources
	  and destinations
		Timing dependent issues
		  Application response
		  time
		  NIC impact
		  Route computation

Impact	Impact of WAN errors on retransmissions and response time	Impact of protocol flow control
Traffic included in simulation	Imported traffic (conversation pairs)	Imported traffic (conversation pairs)
Application types	Standard applications Custom/nonstandard applica- tions	Standard applications Applications from library models
Reports	Recommendations Link-segment utilization	Link-segment utilization
Graphs	Recommendations	Profile (time) data Transit time analysis Utilization on each port of an Ethernet switch

works. However, it cannot be used to predict the effects of several interesting behaviors which transport protocols and applications can exhibit
under nonideal network conditions.

Hence, the need for event-driven simulation—to identify the performance impacts of transport protocols and application behaviors. Table 6-7
compares the two types of simulation.

Vendors of network simulation tools include CACI, Grid Technologies,
Make Systems, Optimal Networks, and Netfusion (formerly Salford Networking). Note that all tools do not support both types of simulation
techniques. For example, CACI's toolset only supports discrete-event simulation for networks.

A Simulation Methodology.   To use simulation effectively, you must
use it in the context of a methodology which also includes collecting
the data needed for simulation. The better the data, the more valid your
simulations. A methodology is needed for predicting the performance
of a network if VLANs were to be rolled out.

You must know both the topology and the baseline traffic of the existing network before building a simulation model. Then, if you are planning to assess the impact of VLANs on the network, it is vital that you understand the type of traffic the VLAN will put on the network—an application profile. Network traffic analysis can help you identify the traffic generated by a particular application. Once you have established the application profile, new nodes can be created or the application added to existing nodes. Fast simulation is used first to determine the overall loading impacts of the VLAN.

Fast simulation will quickly determine if the existing network has sufficient capacity to support the planned VLAN. In addition, fast simulation can be used to make capacity and other adjustments to the network to get the segment and link utilization within design goals. Next, a full simulation can be run to determine how the transport protocols and protocol overheads affect the results.

The Keys to Effective Simulation. Perhaps the most important key to effective network simulation is to have a valid representation of the network topology and traffic. Collecting data from laboratory trials and/or an existing network is essential. Next, an analysis of the current traffic will allow you to focus the simulation on a portion of your network.

Be aware of the limitations of simulation. Event-driven simulation by its nature can be very computationally expensive. The advances in computing speed in the last decade have made simulation a viable tool for predicting the performance of networks. Despite the increase in computing power, it is often necessary to focus on a subset of a network in order to get statistically valid results in a reasonable time. Brute-force attempts at simulating an entire network are destined for failure.

Ensuring Simulation Validity. Here are some specific suggestions that you can follow to help make sure your simulations are valid:

1. *Validate the model against the existing network, prior to simulating VLAN traffic.* It does not matter how accurate the simulator is if the traffic topology does not represent reality. Validation is a continual task—continue to evaluate the data already collected and validate it against the question you are trying to answer with simulation. Remember, simulation always uses an approximation, and you must determine how much detail is sufficient for solving the problem. If the results run counter to your intuition, then run more experiments (either simulation or laboratory tests) to confirm the validity of the results or identify an anomaly.

2. *Capture traffic over appropriate time periods.* Because network modeling tools builds client/server traffic models based on the conversation pairs imported from RMON probes, it is important to capture over an appropriate time period. The imported RMON conversations indicate the average rate of data transfer in each direction—client-to-server and server-to-client—over the measurement interval. Full simulation converts this into a random request and response with an interquery time which will result in the same traffic rates for the conversation. For example, if you observe client/server transactions occurring at an average rate of one per minute, then capturing traffic for several minutes is required in order to capture the average rate. Collecting for only a minute may cause the RMON probe to capture no transactions or several transactions—an unlikely burst.

3. *Run the simulation long enough.* If you do not run the simulation long enough—which is one of the most common mistakes in discrete-event simulation—your statistical results (e.g., average load, average utilization) will not be accurate. How long is long enough? Generally, simulate long enough so that the simulation (*a*) reaches steady state and (*b*) simulates a representative traffic sample after it has reached steady state. A simulation will reach steady state when the queues—such as in NICs, switches, and routers—are filled to their normal operating point. (When a discrete-event simulation starts, all queues are empty.) There is no exact formula for calculating how long a simulation must run to give accurate results. Modeling small transactions on fast segments may require simulating for several seconds (of simulated time, not real time). If interquery times are large for the VLAN, the simulation must run for a longer time period. Look at the interquery times for the VLAN and simulate at least 20 (preferably 100) occurrences of the VLANs being measured and the applications which put significant load on the segments and links of which VLANs extend. For example, if a transaction occurs every 5 s, you'll need to simulate for 100 to 500 s to capture 20 to 100 transactions.

4. *Simulate several different scenarios.* Because of the random nature of user activity, if a network of VLAN is affected by random usage, model several cases: an average case and perhaps an extreme worst case, probably a heavy load likely to occur at the beginning or end of a month or quarter. What is modeled depends on the business functions which the network supports and the type of risk which can be tolerated. Do not assume that there is a single model of network traffic. Make sure that the limiting resource in the network is one which is modeled well by the simulation (see Table 6-8).

TABLE 6-8

Simulators' Ability to Locate Bottlenecks

Limiting resource (bottleneck)	Simulation type	
	Fast	Full
WAN bandwidth	Excellent	Excellent
LAN bandwidth	Excellent	Excellent
Transport protocol	N/A	Excellent
Router throughput	N/A	Good
NIC	N/A	Good
Client speed	N/A	Poor
Server speed	N/A	Poor

Key to table:

Excellent: The results produced by the simulator can be relied on for determining if the resource will limit throughput in the scenario.

Good: The results produced by the simulator are based on measured data. You should compare the test conditions to your particular application to determine whether there is a risk and then decide if alternate performance prediction techniques are justified.

Poor: Because these devices are not modeled in detail, do not rely on the simulator to determine if that resource will limit performance.

N/A: Not computed.

Accuracy of simulation: How accurate are the results? It's the most common question that arises when using simulation as a performance prediction tool. With any combination of tool and methodology, this question cannot be answered without asking more questions. The accuracy which can be expected under various conditions follows in text.

Fast Simulation. If utilization is low—less than common design rules (30 percent for Ethernet, 50 percent for token ring and FDDI, 70 percent for WANs), then the results are quite accurate—and limited only by the accuracy of the traffic data imported. Importantly, in switched networks and VLANs, look out when importing traffic, especially background broadcast and multicast traffic; if it is not imported, it will have to be added to the model manually. If utilization is higher than the 30/50/70 rules of thumb, the protocols and applications will tend to "throttle down," keeping utilization from going to 100 percent. Because fast simulation does not account for these behaviors, it can report utilization greater than 100 percent—obviously incorrect but a useful indicator.

If the load is that high, then the design of the network will need to change to make it work. Fast simulation is good enough to tell if the load is too high. How traffic is captured prior to import can have a great impact on accuracy. If captured on a single segment, the results for that segment will be accurate, yet the results for other segments will lack the traffic which does not flow through the probed segment. For full simulation, if the guidelines for a successful simulation are followed—in particular, capturing the right traffic and running the simulation long enough—then expect very accurate utilization results. However, latency results are more difficult to validate.

Network Response Time (One-Way or Round-Trip)

To determine the accuracy of network latencies, the composition of the response time needs to be determined. If latency is dominated by transmission or queuing time on a LAN or WAN, results will be quite accurate. If the response time is dominated by processing time in a client node, switch, router, or server, then the results will be less accurate. Typical simulation tools allow control of the processing times in nodes, switches, routers, or servers. These values may need tweaking to match the performance delivered by the devices in a specific configuration.

Application Response Time. *Application response time* is generally defined as the response time end users see between the time they begin an operation—for example, by pressing an OK button on a form—and the computer responds—for example, with a screen full of information. To determine application response time, add a ping application to measure node-to-node response time, then use mathematical techniques or laboratory testing to determine the application response time at that network latency.

VLAN Management Complexity

The reason most often given for VLAN implementation is to reduce the cost of handling moves and changes. Since these costs can be quite large, this argument for VLAN implementation is quite seductive.

Many vendors promise that VLAN implementation will result in a vastly increased ability to manage dynamic networks and realize substantial cost savings. This argument is relatively strong for IP networks. Usually, when a user moves to a different subnet, IP addresses must be manually updated in the workstation. This updating process can consume a large amount of time that could be used more productively.

VLANs eliminate this tedious process, since VLAN membership is not tied to a workstation's physical location in the network, enabling moved workstations to retain their original IP addresses and original subnet membership.

While increasingly dynamic networks absorb a lot of the budget of most IT divisions, merely implementing a VLAN implementation is no guarantee that these costs will be minimized. Remember, VLANs inherently add another layer of virtual connectivity which must be managed in conjunction with physical connectivity. This does not deny VLANs cannot reduce the costs of moves and changes—they will, if properly implemented. Simply throwing VLANs at the network is not a solution; organizations must make sure that VLANs do not create more network administration overhead than they save.

Management tools for VLANs must reduce complexity, lower the cost of ownership, and incorporate policy-driven management processes delivering granular performance and service levels. Many features currently offered by a few vendors will eventually become common among all offerings—all VLAN management offerings are converging in terms of functionality. For example, various levels of views, incorporation of monitoring information, and device management tools will be eventually provided by all vendors.

Virtual LANs are broadcast domains that are defined by management software rather than router interfaces. Administrators can use virtual LANs to contain traffic and alleviate some of the problems of a flattened network by restricting the traffic to specific domains.

According to PowerPoint presentations beloved by vendor marketing managers, VLANs reduce the time to support user moves, adds, and changes. Automatic registration of a system allows it to move within the network and maintain its privileges and memberships without further administrator effort. VLAN management tools simplify network configuration, VLAN membership management, and overall monitoring tasks. Administrators can reconfigure networks with a simple click of their mouse, allegedly. Reality is more complex.

Multiple Integrated Views

One key aspect of a robust virtual LAN management platform is information management: tracking the relationships between physical topology, switch ports, VLAN memberships, policy constraints, and service levels.

VLAN topology can be defined as multiple overlays on the physical topology. In addition to capturing relational information at each layer, the multilayer topology system must be capable of correlating the topology model at each layer with the models at the other layers. This allows the network administrator to select a region of the network and transform between topology views to examine it from each available perspective. However, a multilayer topology is much more challenging to implement because now all the other internetworking devices (hubs, switches, bridges) must also be aware of how they are connected to the rest of the network—and, although several VLAN vendors have VLAN management products which derive their views from a common database, the information contained in the database is incomplete. Enhanced awareness of connectivity requires that a high degree of intelligence be incorporated in embedded management agents, as well as in the management application system.

Another basic requirement is a mechanism to allow the agents in each device to exchange topology information with other agents in neighboring devices. Each management agent must then maintain a table of local topology information in an MIB which can be retrieved via SNMP by the system building the global topology. The embedded agents must also be aware of VLANs and include this information in the local topology MIBs.

The global multilayer topology database must have a well-structured database schema to allow tight integration of logical, physical, and virtual views of the network. This requires sophisticated algorithms which are highly optimized for synthesizing information from multiple devices and technologies into a single relational model. In addition, a near-real-time updating mechanism must be provided to allow the database to accurately reflect topology changes when new devices are installed or reconfigured or an existing network element fails.

The virtual view of the multilayer topology database eases graphical applications for configuring VLANs which can span frame-switched, cell-switched, and configuration-switched shared-media networking technologies. Again, the physical-to-virtual mapping of the multilayer topology enables a systemwide autoconfiguration of appropriate interswitch links.

RMON1 and RMON2 probes and embedded agents provide a wealth of real-time and historical traffic statistics at the MAC, network, and application layers. For RMON1- and RMON2-based applications to realize their full potential, it is necessary to be able to map traffic data directly to both the physical and virtual views of the network. Thus, a VLAN management system should provide these views as a minimum:

- Physical topology
- Logical topology
- Device views
- Virtual workgroups on physical topology
- Virtual workgroups in a single view
- Iconized workgroups
- Hierarchical workgroups

Most vendors don't provide these. Multiple, integrated views of the managed environment, including topology, virtual LAN membership, and device-centric-views—which integrate physical and higher-layer levels together, perhaps showing overlaid workgroup icons on a per port basis on device front panel—of emulated LANs across ATM backbones, and views of application- and traffic-related activities, are also needed. These views must be consistently and easily linked to enable management staff to navigate between them. This is a long-unfulfilled list—and yes, VLAN management is still an immature technology and generally not well integrated with vendor network management platforms.

Policy-Based Management

When Novell introduced NetWare 4.x, network administrators found that they were not just installing a new version of an enterprise network operating system. For the first time, many administrators had to organize a standard naming convention for networked resources and establish guidelines for giving users access to those resources. In brief, an upgrade to NetWare 4.x meant creating enterprisewide policies to govern the network for the first time. And, until the recent arrival of tools for Novell NDS (Network Directory Services), this had to be carried out manually.

Since then, the idea of policy-based management has caught on as a way to reduce administrative costs, tighten security, and aid troubleshooting by setting standards for dealing with a multiplicity of situations, ranging from adding new users to the network to reacting to system faults. To facilitate the process of coordinating corporate policies and management practices, network and systems management applications developers have begun to build policy-based management capabilities into their tools.

Many administrators already have written policies in place to help them manage the network. Backing up file servers every Friday evening or running inventory on the network twice a year are examples of common policies used to maintain network operations. The only difference is that these policies aren't automated. Rather than being organized in a database, management policies are often written on scratchpads, stored as email for network operational staff, or as a mental note.

There has to be a clear business case for implementing policy-based management. Policy-based management can be particularly valuable to financial, government, research, and other organizations where access to systems needs to be strictly regulated. Nonetheless, companies need to do a lot of work beforehand to get the most out of policy-based management.

Automated Management

Companies evaluating policy-based network and VLAN management need to examine the administration processes they have in place now and compare that to the processes they want automated. From this starting point, companies can work out whether the policy-based management products—including policy-based VLANs—currently available present a realistic solution.

VLAN management capabilities are software-based. Hardware in the switches carries out the management system directions, but control and decisions are made in software. This software-intensive approach incorporates policies dictating actions of the management system, to manage technology not only for availability or performance but also for business purposes. A range of policies will include aspects of switching and virtual LANs. Policy templates allow administrators to select the parameters to drive the automated management processes.

Policies embedded in VLAN management can be separated into two basic but intertwined types: business and technical policies. *Business policies* include identification of: VLAN availability, service levels, critical applications, and critical network resources; while *technical policies* deal with the nuts and bolts of implementing the higher-level business policies. Technical policy implementation covers such items as default configurations, bandwidth allocations, priorities, VLAN membership changes, security, and quality of service. Both business and technical policies are rules which drive the actions of the VLAN management system. Typical technical policy-based VLAN management allows new users to be dragged and dropped from one VLAN to another with associated policy profiles which include access control to network resources, time-of-day and day-of-week when access is allowed, and the degree of security required by the virtual workgroup or individual user.

Routing Policies

Routing functions will change in the switched VLAN where inter-VLAN connectivity is needed. The traditional routing function is contained in a single device with multiple LAN interfaces. This approach is an interim step to begin building VLANs, and is a speed bump in a switched fabric. The path-selection and packet-forwarding functions must be decoupled in a switched environment. Path selection finds the appropriate path between two points in the switched fabric based on rules and constraints; forwarding uses different techniques to keep up with higher speeds. A mature routing solution has distributed forwarding and path selection in a combination of packages.

Forwarding. Packet forwarding must be distributed across the switching fabric to maintain low latency and avoid the congestion associated with a one-armed router. Distributed forwarding (better known as *layer 3 switching*) minimizes latency by local forwarding. Switches examine the network layer address and forward the packet as instructed by a route-server.

The key factors for layer 3 switching are fast silicon and appropriate placement to minimize the number of hops and a reduction in the complexity and costs of all switches in the network.

Path Selection. The path selection function (or route-server) makes the appropriate routing selections incorporating any metrics for quality-of-service needs and policy requirements. The forwarding information is cached in the appropriate layer 3 switches so that they can make forwarding decisions.

Path selection can be centralized or distributed. Centralized path selection is a transitional step because it does not address issues of scalability or fault tolerance. However, a centralized path selection scheme uses existing technology and may be simpler as initial implementation for a vendor.

The central route-server is a potential single point of failure or a bottleneck within a network. However, it eliminates the one-VLAN-per-router-port relationship created by using an external router.

Distributed path selection is more appropriate for large-scale environments where scalability, latency in route setup, and fault tolerance are essential. Distributed forwarding is more complex. Each path selection element in the network must communicate with its partners to exchange and update reachability information across the switching network. Distributed path selection builds on experience with routing protocols, such as RIP, IGRP, OSPF, and now MOSPF and NHRP.

Intra-VLAN Routing

Routing functions within a VLAN become more important as the geographic spread of the virtual LAN continues. Members of the VLAN are on multiple switches, separated by a mix of backbone technologies. Intra-VLAN routing must accommodate QoS requirements for the VLAN itself. For example, members of a multicast VLAN doing video-conferencing require higher-speed backbone links between members than those doing regular messaging. Intra-VLAN routing requires appropriate path selection based on QoS metrics and the ability to incorporate multiple technologies (e.g., ATM, Ethernet).

Other important features allow traffic control through the use of separate spanning trees for each VLAN. Trunking—assigning a set of parallel links to a VLAN—is a useful feature for managing bandwidth between parts of a VLAN.

Routing policies are already common in VLAN implementations, allowing administrators to assign bandwidth and priorities to specific applications, to connections between VLANs or connections within VLANs. These policies are applied as dynamic conditions arise. For example, for a failed interswitch link, new routing policies are activated to ensure that mission-critical traffic gets the remaining bandwidth.

Access Control

The capability of VLANs to create firewalls can also provide more stringent security, replacing a lot of the functionality of a router. This holds true when VLANs are implemented in conjunction with private port switching. The only broadcast traffic on a single-user segment would be from that user's VLAN (i.e., traffic intended for that user). Conversely, it will be impossible to listen to broadcast or unicast traffic not intended for that user—even by putting the workstation's network adapter in promiscuous mode—because the traffic does not physically traverse that segment.

Access control policies, applied at many levels, usually overlap with membership and configuration policies, as well as define more granular access within a VLAN or across VLANs.

VLAN Configuration Policies

Configuration policies dictate initial settings for switches and topology organization. This ensures that the adaptive switch network is changed to accommodate particular requirements. For example, at night configurations are changed to deny external access to key servers; during daytime, other policies are effective.

A key issue to VLAN deployment is how much VLAN configuration is automated. To a degree, this amount of automation depends on how VLANs are defined. In the final analysis, the specific vendor implementation determines this level of automation. There are three fundamental levels of automation: manual, semiautomated, and fully automatic.

Manual. With purely manual VLAN configuration, both the initial setup and all subsequent moves and changes are controlled by the network administrator. Purely manual configuration gives a high degree of control. However, in large enterprise networks, manual configuration is often impractical. Essentially, it defeats one of the primary benefits of VLANs: elimination of the time it takes to administer moves and changes. Moving users manually with VLANs could actually be easier than moving users across router subnets, though, depending on the vendor's VLAN management interface.

Semiautomated. *Semiautomated configuration* refers to the option to automate either initial configuration, subsequent reconfigurations (moves, changes, adds), or both. Initial configuration automation is usually accomplished with a set of tools mapping VLANs to existing subnets or other criteria. Semiautomated configuration can also refer to VLANs which are initially configured manually but with all subsequent moves being tracked automatically. Combining both initial and subsequent configuration automation still implies semiautomated configuration, because the network administrator always has the option of manual configuration.

Fully automatic. A system which fully automates VLAN configuration implies that workstations automatically and dynamically join VLANs depending on application, user ID, or other criteria or policies determined by the administrator.

Membership Policies

Membership policies define the membership criteria for joining a specific VLAN. Other rules for changing membership may require specific authorization—for example, only a high-level administrator could add or remove members from a closed, private VLAN. Other rules may make constraints about changes based on time of day, network activities, or other parameters. Moving a member between VLANs should also have restrictions placed on who can perform the operation and with what authorization.

Policy-based VLANs—Not Fully Automated

Consider a simple switched backbone with redundant links between switches. If a link fails, there is still connectivity between switches, although the backbone capacity may be dramatically reduced. A policy-based management system ensures that mission-critical activities receive the remaining bandwidth and that essential business activities continue with minimal interruption.

Lower-priority activities receive less bandwidth than normal or may be suspended until adequate capacity is restored. This is a decision an operator sitting at a management console cannot be expected to make—the rules and criteria must be embedded in the management system itself and carried out as automatically as possible. A first step toward fully automated policy management would include displaying instructions and identifying steps the operator must take to ensure that mission-critical traffic continues to flow.

Monitoring

Traffic analysis tools are essential for understanding the distribution of traffic across the physical backbone, traffic within a VLAN or application, or protocol distributions. In addition, monitoring should be fine-grained enough—by virtual workgroup, by resources, automated suggestions, profiling—that administrators can trace an end-to-end path between pairs of systems and understand the behavior, congestion, and other conditions on each leg of that path.

Modeling and Simulation

Modeling and simulation tools provide information for optimum placement of clients, servers, backbone links, and other resources in the network. In a similar way, modeling tools help administrators predict the impacts of changes. For example, administrators need to answer questions such as

What happens when I add 200 more Lotus Notes users?

What happens when 50 users are moved from one building to another?

How does that affect backbone traffic, loading, and server access patterns?

Modeling tools, both tactical and strategic, use data collected with the monitoring system to drive models and calibrate results. Tactical tools solve short-term problems caused by temporary congestion or failures. Strategic utilization of modeling tools projects long-term changes, analyzes several topology and placement alternatives, and runs different traffic scenarios to ensure that growth will not compromise enterprise requirements.

Capacity planning tools take collected information and automatically generate trends identifying points where resource limitations will occur unless corrective action is taken. Planning tools should be automated to update projections on an ongoing basis.

A Simple Move?

Theoretically, a person on a VLAN can move to a new physical location but remain in the same department without having workstations reconfigured. This also implies that a user's physical location does not necessarily change when the user transfers from one department to another—the network administrator simply changes the user's VLAN membership.

From a network management perspective, the transitory nature of these virtual workgroups may grow to the point where updating VLAN membership becomes as tedious as updating routing tables to keep up with adds, moves, and changes today, despite saving time and effort in physically moving the user's workstation. In addition, there are social hurdles to overcome in the virtual workgroup—people in teams usually like to be physically close to those with whom they work, rather than for the technically attractive reason of reducing traffic across a collapsed backbone.

Virtual LAN support for virtual workgroups is often tied to support of the 80/20 rule (80 percent of the traffic passes between workgroup members while 20 percent is remote from or outside the workgroup). Theoretically, by properly configuring VLANs to match workgroups, only the 20 percent of the traffic that is remote needs to pass through a wide area router and out of the workgroup. Consequently, this improves performance for the 80 percent of the traffic that is within the workgroup.

However, the 80/20 rule is under threat from the deployment of servers and network applications such as email, Lotus Notes, and intranets/Web browsers which supplement the traditional user-to-server traffic with user-to-user and server-to-server traffic across the enterprise.

The virtual workgroup may simply run into problems such as requiring that users sometimes be physically close to certain resources, such as printers. For example, a user may be logically in the accounting VLAN but physically located in an area surrounded by members of the sales VLAN. The local network printer will also be in the sales VLAN. Every time this accounting VLAN member prints to the local printer, this member's print file must traverse a router connecting the two VLANs. This problem can be avoided by making that printer a member of both VLANs. This example emphasizes VLAN implementations which provide for overlapping VLANs. If overlapping VLANs are impossible, then the routing functionality must be built into the backbone switch. Thus, the print file would be routed by the switch rather than by an external router.

Server farms are a return to the "glasshouse" with departmental servers located in a centralized datacenter, where they can be provided with consolidated backup, uninterrupted power supply (UPS), and environmental protection. Centralized server farms create problems for the virtual workgroup when vendors cannot allow a server to belong to more than one VLAN simultaneously. If overlapping VLANs are not possible, traffic between a centralized server and clients not belonging to that server's VLAN must traverse a router. However, if the switch incorporates integral routing and is able to route inter-VLAN packets at wire speed, there is no performance advantage for overlapping VLANs over routing between VLANs to allow universal access to a centralized server. In this scenario, only inter-VLAN packets would need to be routed—not all packets. Several vendors provide integrated (layer 3) routing as an alternative to overlapping VLANs. While workgroup VLANs may be extended to centralized server farms—for example, including a particular fileserver in a particular workgroup's VLAN—this is not always possible.

Policy layers must be organized. Mobile users may require a temporary change in policy. For example, visitors from other parts of the enterprise should be able to sit down at any computer, properly identify themselves, and have their normal virtual LAN access to resources, while other members of the VLAN are constrained by normal policies.

Security management, for example, requires policies to determine which users have access to what network resources. Without such policies, security is bound to be compromised. Policies can also help streamline administration tasks, such as configuring new users' desktops and preventing the use of unauthorized software.

For policy-based VLANs, even a simple move becomes a complex operation, hidden by the degree of automation of the policy process. Thus, a move may require

- Policy oversight
- New workgroup software
- New workgroup virtual circuits
- New access privileges
- Access to new servers

Virtual Management Integration

Comprehensive VLAN management applications provide access—possibly through a set of logical and physical APIs—to other applications tools in the network management environment. For example, an inventory tool needs access to both physical and logical information. The logical information, at this point, resides within the VLAN management system and must be visible to other tools. An inventory management request might ask "How many 486 processors are in the sales VLAN?" Or it may ask for the number of Pentium Pro servers in the enterprise, regardless of VLAN membership.

Similarly, a software distribution tool could pick as a target a logical group, such as finance or marketing VLANs, or may work in the traditional way, with a specific set of target systems. Making the VLAN information available to other management tools helps reduce the cost of ownership, simplify management processes and procedures, and reduce complexity for the management staff. Unfortunately, as yet, no VLAN vendor has published its VLAN API—they may not exist.

Integration with Network Management and Other Management Tools. Switching coexists with the installed base of shared media LANs and wide area backbone technologies. Information from the VLAN management system must be integrated with the network management environment to provide administrators with a complete view of the behavior of switched networks in relation to shared media and WAN technologies. This has yet to be addressed by most VLAN vendors.

Integration with other management tools is also essential. For instance, information from the VLAN management system must be provided to help-desk applications to provide the staff with up-to-date information about VLAN membership, policy restriction, and resources when a user calls with a problem. In the same way, trouble-ticketing systems must be integrated so that VLAN management tools automatically create and populate a trouble ticket and allow staff to track a problem through to resolution.

Policy-Setting Alternatives

Policy setting by VLAN management is only one method of handing out policies to a network. Currently, the only working alternative is for policies to be set by the network management system. A third possibility, the first example of which is likely to be delivered by Cisco and 3Com by the end of 1997, is for policies to be handed out and enforced by a network directory services application.

Thus, policy for a network can be set by

- VLAN management
- Network management
- Directory services

Network Management Applications

Policy-based management is also making its way into high-end network management applications. Hewlett-Packard has an add-on to its HP OpenView platform, called *AdminCenter.* The UNIX-driven management software is aimed at streamlining administrative jobs such as configuring users and applications. But unlike current VLAN policy-based solutions, AdminCenter provides a sophisticated mechanism for organizing policies and reconciling conflicting policies.

The software uses a knowledge base to store the policies hierarchically. Administrators can designate policies for the global environment, for specific domains, or even for specific workstations. AdminCenter automatically assigns ratings to policies that allow very specific, low-level policies to overrule more general, high-level policies.

For example, a high-level policy dictating that only users with an ID number between 100 and 500 can access the accounting server may be overruled by a low-level policy which says that users in the remote accounting office can also access the server. But high-level policies are overruled only if the administrator who writes the low-level policy has the appropriate access privileges.

Directory Services

Computer and network administration is incredibly messy. The point nature of IT security means that you must currently separately maintain router tables, router path management, firewall filters, and access server privileges—all with separate management tools.

Ideally, a directory service provides a single place to register all IT and network assets in an organization and makes the process for doing so formal, controlled, and often automated. Once registered, these assets can be tracked and managed more easily than the myriad of user access control lists, VLAN memberships, domains, and rights managers in use today.

Once established, the directory becomes a database of all devices and specific attributes on the network and the relationship between those devices. It also becomes available for use by other services—including security services—users, applications, and processes running in the network.

The multiplicity of policies and policy-setters gives rise to several practical questions:

Who sets policy for the VLAN?

Is there a hierarchy of policies in operation?

Do the different policy-setting engines compete or cooperate?

How are conflicting policies resolved?

Will vendors' directory services be the source of their policy-based
   VLANs, or will it be embedded in a VLAN management application
   or its network management application?

Unfortunately, it may be some time before answers surface—and decide what policy-setter is best for which VLAN.

Service-Based VLANs

Strategically, these organizations can deploy VLANs as either an infrastructural or a service-based VLAN implementation. The choice of approach will have a substantial impact on the overall network architecture which may affect the management structure and its business.

Infrastructural VLANs. An infrastructural approach to VLANs is based on the functional groups (the departments, workgroups, sections, etc.) making up the organization. Each functional group, such as accounting, sales, and engineering, is assigned to its own uniquely defined VLAN. According to the 80/20 rule, the majority of network traffic is assumed to be within these functional groups, and thus within each VLAN. In this model, VLAN overlap occurs at junctures where network resources must be shared by multiple workgroups. These resources are mostly servers, but could also include printers, routers providing WAN access, and workstations acting as gateways.

The amount of VLAN overlap in the infrastructural model is minimal, involving only servers rather than user workstations, thus easing VLAN administration. Generally, this approach fits well in those organizations with clean discrete organizational boundaries. The infrastructural model is also the approach most easily developed from current VLAN implementations and is a good fit to current networks. Further, the infrastructural approach does not require network administrators to alter how they view the network and also means a lower cost of deployment. For these reasons, most organizations should start with an infrastructural approach to VLAN implementation.

Service-Based VLANs. A service-based approach to VLAN implementation looks not at organizational or functional groups but at individual user access to servers and applications, i.e., network resources. In this model, each VLAN corresponds to a server or service on the network. Servers do not belong to multiple VLANs—groups of users do. In a typical organization, all users would belong to the email server's VLAN, while only a specified group such as the accounting department plus top-level executives would be members of the accounting database server's VLAN.

The service-based approach inherently creates a much more complex set of VLAN membership relationships to be managed. With the immaturity of most current VLAN visualization tools, a large number of overlapping VLANs using the service-based approach could generate incomprehensible multilevel network diagrams at a management console. Consequently, service-based VLAN solutions must include a high level of automatic configuration features. However, in response to the types of applications organizations want to deploy in the future, together with the shift away from traditional, more hierarchical organizational structures, the trend in VLAN implementation is toward the service-based approach.

As bandwidth to the desktop increases and as vendor solutions become available to better manager greater VLAN overlap, the size of the groups belonging to a particular set of VLANs may become smaller and smaller. At the same time, the number of these groups becomes larger and larger, to the point where all users might have a customized mix of services delivered to their workstations. Taking this a step further, control over what services are delivered at a given time could be left up to each individual user. At that point, the network structure begins to look like a multichannel broadcast network. In fact, by this stage, the service-based model finds the greatest affinity in VLANs defined by the IP multicast group—each workstation has the choice of which IP multicast or channel it wants to belong to.

In such a future, VLANs lose the characteristics of static or semistatic broadcast domains defined by the network manager, becoming channels to which users subscribe. Users sign up for the applications they need delivered to them at a particular time. Application use could be accounted for, enabling precise and automated chargeback for network services. Network managers could also retain control in order to block access to specific channels by certain users for security purposes.

As vendor VLAN offerings develop, many organizations will want to consider migration toward a service-based model, which will let users subscribe to various network services. This concept of user-controlled subscription, compared to administrator-controlled membership, is aided by NICs with built-in VLAN functionality operating in environments with a single user per switch port.

In this scheme, the NIC driver dynamically tells the switch which multicast groups or VLANs it wants to belong to. This type of distributed VLAN control takes advantage of the increasing processing power of the desktop and provides a higher degree of related functionality such as automatic VLAN configuration and traffic monitoring. In addition, agents in each NIC will enable the workstation to collect and report information on specific application usage (rather than just simple layer 2 traffic statistics in the case of RMON1). This capability will make automated chargeback for network services for service-based VLANs much easier. This seems to be the next stage of evolution for 3Com's and Cisco's VLANs.

For example, 3Com's VLAN management application is *ATMVLAN Manager*, which is part of Transcend Enterprise Manager for UNIX. Integration between the this and Transcend Enterprise Manager is tight, including common launch, icons, graphical user interface (GUI), and database. Integration between the Transcend Enterprise Manager for UNIX and the underlying SNMP platform is not as tight as between 3Com applications. Event management, menus, and MIB integration are available, but there is no shared database. However, 3Com is developing policy-based management to leverage their SmartAgents. Initial efforts are focused on policy-based security to define access rights to user by time-of-day and other criteria. 3Com is also making efforts toward service-based VLANs, although these are also weakly integrated with its VLAN management toolset.

3Com intends to add InfoVista's service-level agreement (SLA) conformance management system to its TranscendWare set of modules, enabling it to enforce SLA metrics (e.g., network uptime, application availability, and network and application response time). InfoVista will apparently enable set, manage, and enforce network policies. However, because 3Com still lacks a common database for its toolset, there is likely to be conflict between network policy set by the InfoVista modules and 3Com's TranscendWare VLAN management tools.

If individual users control VLAN membership, what about security? Clearly, users cannot be allowed to subscribe to any network service they wish. The network administrator must be able to establish policies to define which users have access to what resources and what class of service each user is entitled to. One solution to the security problem may come in the form of an authentication server. These servers could develop into the main method of defining future VLANs.

Authentication servers define VLAN membership by user ID—a password or other authentication device—rather than by MAC address or IP address. Defining VLANs in this way greatly increases flexibility. It also implies a certain level of integration of VLANs with the network operating system, which usually asks the user for a password to allow or deny access to network resources. One of the main advantages of authentication servers is that they allow users to take their own VLANs anywhere, without regard to which workstation or protocol is in use. At the moment, VLANs have great difficulty in dealing with the management issues associated with remote and mobile users. In these cases, the assignment of an IP address to a VLAN user would occur only after authentication. Currently, we are awaiting the arrival of a security server which can assign VLAN memberships (and therefore security policy) rather than simply assigning access privileges to system resources.

Policy-Based VLANs

As is becoming apparent, the integration of policy-based management is in its early days—with a narrow technical scope—and the necessary VLAN management hardware needs to be embedded in switches. Ideally, VLANs are created and managed with software which allows for incorporation of rules and restrictions to control and direct the management system. Management tasks are carried out to satisfy policy rules supporting the business processes of each customer. For instance, a policy identifies critical applications and ensures that users continue to receive the bandwidth they need when failures occur. For a successful enterprisewide VLAN implementation, modeling and analysis for the placement of switches is critical.

You should ask your VLAN vendor what policies (if any) are supported—and about future development directions. Policy-based VLANs differ tremendously in the width and depth of their implementation and the granularity of the individual component policies:

- What management policies are available, and what is their scope?
- What management templates are supplied?
- What are the precedence rules for conflicting policies?
- How are moves, adds, and changes dealt with?
- Do policies "follow" the user, or is a new policy required for a move?
- What configuration policies exist?
- Does the policy cover bandwidth allocation and prioritization?
- How is traffic management integrated into bandwidth management policy?
- How sophisticated is the VLAN security policy?

Traffic Management

When VLANs are deployed in existing networks, they can affect the performance and reliability of other applications in ways which are difficult to predict, understand, isolate, and solve. Problems can range from graceful degradation, to intermittent failure, to disaster. If the network infrastructure moves from router-based to switch-based, this compounds problems as network support staff gradually familiarize themselves with new technology.

Before considering costly upgrades to workgroup server performance or increasing the bandwidth across the network backbone to support switches and VLANs, first examine the existing network to discover why problems occur and how to solve them—this is also the starting point for optimizing network performance. Monitoring your network will help you

- Identify traffic flow patterns
- Identify congested segments of the network

It will also help you design the future switched network, determine your bandwidth requirements (and bandwidth trends), determine what size of switch will generally meet which traffic demands, and decide where switches should be placed first in order to relieve potential and actual network congestion problems, in a phased implementation of switching technology and virtual LANs.

The Ugly Side of VLANs

The traffic management side of VLANs is not pretty. A clever protocol, SNMP, dominates today's network management platforms for distributed devices. Unfortunately, SNMP was developed to manage physical network devices—not VLANs. It is not a simple task to retrofit the virtualization of LANs onto SNMP, a determinedly nonvirtual protocol.

SNMP's developers recognized its shortcomings, and responded with the Remote NETwork Monitoring (RMON) protocol. RMON looks at LAN segments—this is the last piece of good news you're going to hear for a while—and can monitor the health of groups of network devices. Unfortunately, what RMON needs is the ability to dynamically manage each virtual port of a virtual physical-layer switch. In terms of processor usage, this is not cheap and it is not easy—RMON is capable of crippling the fastest switch, whether Ethernet, token ring, or ATM. VLAN concentrators are best at moving packets, not analyzing them.

Traffic management for ATM is even more gruesome. Using SNMP to manage ATM is like using an abacus to solve a linear programming equation. Today's SNMP platforms have problems managing hundreds of events in large networks—never mind correlating those network events in a meaningful manner—we live in hope. Remember, you cannot manage what you cannot measure.

The ATM Forum has only begun to deal with the congestion control and ATM's AToM MIB integration with SNMP (see preceding discussion). On the ATM side, the good news is that the ATM Forum is tackling a few minor pieces of the puzzle. The bad news is this is still an interim plan which addresses neither end-to-end delivery nor the lack of real-time capabilities in SNMP. More bad news is that, as yet, there is no agreement on an RMON management information base (MIB) for ATM.

In 1996, Cisco announced that it was working with AXON (now part of 3Com), Frontier, Net2Net, and Network General to develop such an MIB. The ATM MIB allows network managers to collect statistics on the number of ATM cells transmitted, received, received in error, and dropped as a result of congestion. The proposed standard will also identify different methods for collected traffic information, including the use of circuit steering to directly selected ATM virtual connections to probes or analyzers. The latter is critical since Cisco has not endorsed embedded RMON in its high-end 7000 Series routers because of the adverse performance impact on the processor. Epilogue Technology is still looking at embedding RMON into ATM switch ASICs to address the issues of performance and real-time utilization of the statistics.

As with most network management tools for switched networks and VLANs, vendors and vendor consortiums are announcing separate traffic management strategies and products—some based on third-party products, such as Frontier Software. Once again, the quality and extent of traffic management software you get is dependent on the switches you purchase. It is not unusual to find different traffic management tools for ATM and Ethernet switches, each producing totally different sets of statistics which are difficult, if not impossible, to correlate in any meaningful fashion.

The Remote Network Monitoring Protocol (RMOM) was developed to look at physical network segments, not logical segments. To develop statistics on multiple virtual LANs, LANNET (now part of Madge Networks) developed a switched RMON implementation. Originally, it allowed network administrators to look at statistics across multiple virtual segments on Ethernet switches, and if a segment has a problem, then the RMON statistics features can look for more granular information.

The Madge SMONMaster 2.0 application extends that capability to include Madge Collage 740 ATM switches and their cell-based traffic as well as Madge Ethernet LANswitches from the same console. With the SMONMaster, network managers will be able to integrate ATM equipment into Ethernet-switched LAN environments, reducing management complexity and reducing costs associated with supporting two dissimilar network management environments (ATM and switched Ethernet).

The SMONMaster application also monitors ATM and switching services such as VLANs (virtual LANs), ELANs (Emulated LANs), signaling, and QoS traffic. The application also measures overall switch utilization. For ATM switches, the application measures overall QoS allocations. It spots anomalies such as excessive signaling or higher-than-normal error rates. It also lets network managers study trends and analyze network activity. However, the integration is only at the console level, not at the functional, collaborative level, but at least it sets a marker for other vendors—in terms of end-to-end switch traffic visibility and integration. That's the good news (but only for those with Madge switches and switching hubs).

There are still problems. First, how much data is there, and how often do you collect it? Once collected, how do you send the information efficiently to one or more management consoles? The problem is the lack of bulk transmission in SNMP which makes the RMON-to-management console transfer a bottleneck.

It should now be apparent that no one vendor has the total solution, but no two vendors can guarantee interoperability. Before you purchase virtual LAN switches and/or ATM switches and/or traffic management products, ask these questions:

■ How do I address fault, performance, and configuration management within the VLAN?

■ Is RMON addressed by the VLAN switch vendor? Is it proprietary?

Traffic management deals with keeping the network up and running. To ensure that this happens, network monitoring and analysis applications are deployed to detect, diagnose, and isolate potential problems on the network. Traffic management tools offer a variety of capabilities, such as baselining the "normal" behavior of the network, analyzing network utilization and trends, identifying performance bottlenecks, and tracing routing paths to identify specific loading of multiple segments.

Tracking trends in a network enables you to anticipate resources problems before they become serious and plan a network redesign or upgrade the critical network element. Used in harness with modeling tools, the information can help you understand the impact on a network of switch and server placements as a prelude to a move toward virtual LANs. In conjunction with a forward-looking network management strategy, traffic management tools will enable VLANs to be planned and implemented while optimizing performance across the enterprise network.

The RMON standard underpins traffic management in networks. Whether implemented as standalone hardware probe or embedded into network devices such as switches, it provides an easy way to monitor layers 1 and 2 of the OSI model. RMON's strengths lie in its ability to provide segment statistics, long-term trending, node traffic pattern analysis, top bandwidth users, alarm reporting, and packet capture. It is a useful tool to identify who caused a network problem and when it happened.

But RMON also has its limitations. While it tells who caused the problem and when, it gives no indication of what the problem is, or was, or why it happened. More importantly, RMON1 addresses only the physical layers of the network—the basic connectivity of cables, switches, and related devices. However, RMON2 addresses these deficiencies in some respect, but until it is widely implemented in switches, there is the danger of relying exclusively on RMON information for VLAN information. Neither RMON1 nor RMON2 addresses VLAN implementations, and RMON vendors have been forced to add proprietary extensions in their implementations to provide VLAN monitoring.

Because of the lower-layer management statistics provided by RMON, the conclusions that the statistics inevitably lead to in order to avoid a performance problem is to either increase the network bandwidth or upgrade key server hardware. This narrowly focused approach can be costly, time-consuming, and sometimes misleading. A better approach for deploying VLANs is to identify which segments on the network are intended to support VLANs. Since a VLAN will include specific workgroup segments as well as the network backbone, traffic monitoring and analysis tools should be deployed on those segments.

To complement these tools, RMON agents and probes can be cost-effectively deployed on several local and remote segments together with embedded RMON within switches to provide continuous operation and information on lower-layer statistics and alarms.

Monitoring

Monitoring the switched environment is a key requirement to provide the information needed for effective management and administration. Most monitoring today is centered around RMON1 agents embedded throughout the network in switches as well as standalone probes. Vendors must move toward a full-scale monitoring system rather than offering a set of point collection devices, whether embedded or standalone.

Embedded RMON1 Agents. Basic per port monitoring is straightforward with silicon implementations of the four basic RMON1 groups: Statistics, History, Alarms, and Events. This level of monitoring must be considered basic to the existence of a monitoring story at all. Embedded monitoring must be available in all switching products to provide consistent and accurate coverage.

The other RMON1 groups, especially filtering and capturing, require more processing and memory and may impact switching performance. Alternatives for providing deeper analysis include roving monitors, external monitors, and steering. Embedding additional monitoring groups in the switch gives a roving probe which brings full RMON1 to any port selected by the administrator.

External Agents. Delivering selected traffic from a switch port to an external collector is another way to provide more detailed monitoring functions. An attached network monitor or protocol analyzer collects detailed information for analysis. This information can be used at the switch or delivered to a remote management center.

Selected port traffic can be "steered" across the network to a remote collection center. Steering has an advantage over a diagnostic port—it reduces the number of expensive monitoring devices and reduces time for troubleshooting since it delivers traffic for analysis quickly and automatically.

Other factors must be considered with steering. For example, delivering streams from multiple switches may add congestion on critical backbones. Another concern is forwarding corrupted packets through a store-and-forward switching network where they are discarded by intermediate switches. Steering is a tool for reactive rather than proactive management.

Full RMON1 per Port. Full RMON1 capability per port can be made by a switch vendor using any of the solutions listed above. But full (nine-group) simultaneous RMON on each port is a rarity today because of the resource demands on the switch or the need for multiple external agents. Simultaneous monitoring improves troubleshooting and offers detailed examination of switching behavior.

High-Speed Monitoring. Interswitch links are rapidly moving toward Fast Ethernet—and eventually Gigabit Ethernet—and the use of 100 Mbits/s switching is accelerating. Currently, embedded monitors cannot capture information at these speeds without serious switch performance degradation.

External monitors (standalone RMON probes) must be attached to interswitch links, high-performance server ports, and other high-speed parts of the switching backbone. The key for these high-speed devices is easy attachment and detachment without requiring interruption of cabling.

Most high-speed monitoring depends on external agents because of the higher resource demands of wire-speed capture. The ability to attach an external high-speed monitor and mirror traffic is necessary.

The key components for a robust and acceptable monitoring solution include

- The basic four RMON groups (the absolute minimum—seven is better)
- A roving agent, steering traffic, or an external monitor
- Simultaneous RMON for each port

RMON2. Embedding RMON2 in switches is necessary to view flows between parts of a switched network, subnetworks, or VLANs in the future. It also provides basic information related to protocol, usage distributions, and application activities. It is expected to appear first in backbone switches; many vendors have yet to announce the scale of their support for the emerging RMON2 standard.

Monitoring Applications: Traffic Flows. Applications exploiting the information collected by embedded and standalone monitors is the key differentiator for monitoring solutions. Applications must collect, integrate, and correlate information from many sources: embedded agents and switches, monitors attached to diagnostic ports, standalone probes on high-speed links, and others.

One key tool is an overall view of traffic flows within a switch and between switches. Traffic distributions are necessary to understand behavior within each switch as well as potential stress points in backbones. Switched workgroups span multiple switches, and traffic is isolated through filters and access control mechanisms. Monitoring tools must correlate information to show flows within and between VLANs.

Monitoring applications must also provide detailed information concerning overall protocol distribution. Knowing the distribution and volume of IP, IPX, VINES, and other protocols across the switched environment is important for understanding the behavior of user communities and their overall impact on the physical infrastructure.

Detailed information about application traffic flows, the communities of clients and servers communicating with each other, the traffic volumes, and the patterns of flow are important for performance optimization as well as overall planning. Information from protocol distribution and applications can be used to ensure that information, clients, and servers are appropriately placed to offer highest-speed switching and minimal backbone congestion. This information is also important for modeling the impacts of moves, adds, and changes. Administrators will have warning of potential problems.

Collaborative Monitoring. Distributing RMON and data collection intelligence to critical parts of the network is one part of the strategy for implementing VLANs. Information from these collectors is passed to network management platforms for action by the network manager or administrator. In addition, they can perform more detailed analysis on longer-term network traffic and create reports to reveal what constitutes "normal" network traffic patterns, monitor the health of the network, identify congestion patterns, and establish a network performance baseline.

In summary, collaborative monitoring requires

- Basic port statistics ("mini-RMON") in all switches
- Workgroup switches need these embedded RMON groups:
    —Packet capture
    —Filtering
    —Matrix to end nodes
- Backbone switches need
    —Trunk statistics
    —Matrix to other switches

Distributed Monitoring. The traditional view of RMON1 as a monolithic, nine-group entity is under pressure as monitoring is physically distributed throughout switched networks. Different RMON1 capabilities will be placed in the appropriate switches to provide overall optimization in collecting data with minimal performance impacts. In addition, this makes it easier to correlate key traffic information from multiple sources and potentially provide a deeper analysis of traffic flows within a switched environment.

In the case of workgroup VLANs based on Ethernet switches crossing an ATM backbone, the capture and correlation of selected traffic—such as individual packets rather than random ATM cells—across the switched network would enable centralization of traffic flow monitoring on a single console together with appropriate visual indications of potential and actual network congestion points. Currently, this is more in the discussion stage than in delivered-product stage. 3Com, however, has been shipping distributed RMON (dRMON) on all its NICs, which, when combined with an edge monitor, gives nine-group information at a reasonable price.

Filtering and capture may be more effectively carried out in workgroup switches at the edge of the network closest to where packets originate or terminate. Similarly, high-speed backbone switches are characterized by fewer ports but higher traffic volumes. The most important statistics are traffic matrices and the TopN measurements to define backbone activity and traffic patterns. With the exception of 3Com, the cooperative approach to monitoring is noticeable by its absence from other vendors' agendas.

Baselining

Good monitoring applications supply automatic activity baselines by taking continuous measurements in the background. Baselining applications must be integrated with other parts of the management system so that appropriate utilization and activity thresholds are set to reflect normal behavior. More sophisticated baselining is necessary to show server activities and, ultimately, applications activity, across the switching environment.

However, setting baselines for a switched environment is currently very difficult, given the opaqueness of switches, whether cell- or frame-based, to traffic measurement. Traffic measurement and monitoring is also not well integrated either as a component of switch and VLAN vendors' management suites or, within the suites, the integration of ATM and non-ATM switch traffic measurement and management applications.

A switched environment requires end-to-end traffic monitoring—after all, this was the norm with shared networks, so why should less be expected from a switched network? With switched networks, traffic metrics need to be redefined or perhaps refined, now that a typical switched network may be located across multiple sites, with differing speeds—10 Mbits/s switched and shared Ethernet in the workgroup with Fast Ethernet uplinks to an ATM backbone together with relatively slow wide area links.

In the shared LAN environment, network utilization was a simple, easily understood measurement—witness, in the past, the brief war between shared Ethernet and the original IBM Token Ring in which network utilization levels became part of their marketing. What does network utilization actually mean—if it means anything—in a switched environment? Does it mean utilization of a switch backplane? Raw utilization of ATM bandwidth or utilization of an ATM class of service? The tools are not here. Fundamentally, if you cannot measure it, you cannot manage it, whatever switch vendors say.

What is required—what many vendors have yet to articulate—the provision of end-to-end monitoring to provide consistent meaningful network traffic metrics across shared and switched LANs together with ATM switches. For example, how does the number of ATM cells dropped by a virtual circuit relate to the number of retries by an Ethernet hub in the same network? Can these be related to VLAN performance? Nobody knows, yet.

Setting Baselines and Thresholds. The basic idea of setting a baseline is to capture a "typical" profile of traffic representing "normal" behavior of a device or on a LAN segment. Once the baseline is established, then thresholds for alarms can be set, depending on the baseline.

Baselines should be set for all LAN segments and measurements taken at regular intervals to spot any anomalies (such as rising level of link utilization) arising from a change in network design or the rollout of a new application. In addition, the input/output (I/O) levels of key servers and their activity profile (through Microsoft SMS or other DMI-compliant utility which provides appropriate statistics) should be closely monitored to provide early warning of potential bottlenecks. Once the data has been gathered, thresholds for SNMP/RMON or network management alarms can be set. To set thresholds correctly, a basic understanding of statistics is required.

Normal Distribution and Standard Deviation. In this discussion, it is assumed that events occur totally at random (i.e., that a prior event has no effect on the outcome of the next event). It is further assumed that the distribution of values measured approximates the standard bell curve and that the values measured have a wide normal distribution.

The normal distribution curve is always bell-shaped with a single peak, on either side of which there are equal frequencies of values. The arithmetic mean (or average) will give you the central peak (or baseline), while the standard deviation measures the dispersion of data about the peak. In addition, knowledge of the standard error (or systemic error) will help you determine how many measurements are needed before a threshold can be established with confidence.

The actual shape of the "bell" of a normal curve is determined by the standard deviation of its distribution; once the average and the standard deviation are known, the complete distribution can be plotted. Also note that the arithmetic mean may not be the actual average—it depends on whether the number of values is high enough to approximate all values taken now and in the future. For this reason, the number of samples taken has to be balanced against how deeply you want to "see" into a specific segment or device in order to establish a workable baseline.

If standard deviations are drawn at intervals of one standard deviation (SD) from the mean, then

- A threshold set at ñ1 SD of mean results in about **68 percent** (actually **68.27 percent**) of the values will not create an alarm.

- A threshold set at ñ2 SD of mean results in about **95 percent** (actually **95.45 percent**) of the values will not create an alarm.

- A threshold set at ñ3 SD of mean results in about **99 percent** (actually **99.73 percent**) of values will not create an alarm.

Because of the relationship between the arithmetic mean and standard deviation, the standard deviation is sometimes referred to as a *sigma* (σ), after the Greek letter often used as a summation in math textbooks.

Initially, the threshold should be set at $3\ \sigma$ in order to minimize network traffic from SNMP/RMON probes and to ensure that only abnormal events trigger an alarm. At $3\ \sigma$, the chance of an abnormal event occurring is about 2 in 10,000.

Standard Error. By increasing the number of values sampled, then standard error (systemic error) can be reduced. Standard error does not correspond directly to the number of values because the standard error is a square root. If the number of values is doubled, the standard error is not cut in half. In fact, to halve the standard error, the number of values

TABLE 6-9

Relationship
between Sample
Size and Standard
Error as Sigmas

Sample size	Standard error % (1 σ)	Standard error % (2 σ)	Standard error % (3 σ)
1000	1.37	2.74	4.11
2000	0.97	1.94	2.91
4000	0.68	1.36	2.04
8000	0.48	0.96	1.44

must be increased four times in order to have the standard error (see Table 6-9).

In summary, good advice is to

▪ Take at least 8000 values for the interval to minimize systematic error.

▪ Use arithmetic means (averages) for each interval.

▪ Use standard deviations to calculate thresholds:

—1 σ—68.27 percent confidence

—2 σ—95.45 percent confidence

—3 σ—99.98 percent confidence

▪ Standard deviations are for setting thresholds—outside 3 σ gives about a 2 in 10,000 chance an event as normal.

▪ Set the threshold value at 3 σ or higher initially to detect abnormal values (events).

Over time, you may need to revisit the device or segment to move the baseline upward or downward in the light of data accumulated over time. However, once the first set of baselines has been established, you will need to look at the data and trends to catch anomalies (unexpected major peaks and troughs) in traffic patterns which occur in a weekly and/or monthly cycle and see if they can be "smoothed out" to reduce the stress on the network.

Most traffic management tools establish a baseline automatically. If the tools have some intelligence, then they will recalculate baselines (and thresholds) automatically as they average the incoming data and converge on the actual baseline of the segment or device. Usually, you can tell the tool to recalculate the associated thresholds after a certain period. In this way, your thresholds are not static but variable, responding to prevailing network traffic. And, a final word on thresholds—consider setting a low watermark as well as a high watermark for traffic levels; a segment or device not generating enough traffic to be considered "normal" may be the first sign of a larger problem.

Summary

The ideal traffic management system for VLANs should

- Locate problems, identify unique VLAN characteristics, and automatically make recommendations on how to resolve problems.
- Provide internetwork bandwidth statistics which indicate utilization based on LAN protocols, VLANs, and end users.
- Automatically associate routing paths both intra- and inter-VLAN and learn WAN bandwidth usage patterns.
- Provide reports which include historical trend analysis of peak network performance and network management reports to enable forward planning and expansion.
- Integrate with the current network management platform (Sun's Solstice SunNet Manager, HP OpenView, IBM NetView for AIX).
- Extract data from multiple sources and correlate them logically for a complete end-to-end view of user traffic patterns in VLANs and all the devices a VLAN traverses.
- Make sure that VLANs run effectively over the network at all times.

Security

Switched Networks and VLAN Security

Networks continue at the core of an organization's drive to increase its competitive edge and are deployed to support the company's central missions. As such, networks are rapidly becoming valuable assets as important strategic investments. Valuable assets need protection, and corporate managers must take prudent steps (or be prodded by their auditors) to protect these assets. The security of the network infrastructure itself, then, is an issue (see Fig. 7-1).

Figure 7-1
Security—the enabler

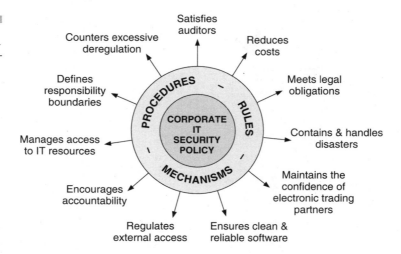

As the infrastructure becomes more complex, the individual components become less a collection of communications hardware, and more a complex interconnection of software-intensive intelligent processors. In short, someone can bring down the network with a change to the software and its controlling parameters. The management and self-management of the network must be protected, or the network can be deliberately (or accidentally) brought down.

Network security is more difficult to manage and control than in the past. Telecommunications deregulation is bringing with it numerous technological options, each with compelling revenue-generating and cost-saving opportunities. Many of these opportunities can be realized only in publicly accessible networks, which increases the list of potential vulnerabilities as well as the scale of the overall threat to the environment. At the same time, the user's ability to set security policies for these public networks is very limited. In addition, commercial pressures are bringing together supplier and customer networks to increase competitive advantages. The fact that these enterprises may have different security policies, or that they might contain business units in direct competition, also compounds the problem.

Open Systems Are Double-Edged

Open system architectures are double-edged swords. They allow businesses to bring together various systems and technologies and quickly integrate them in a way that was impossible under previous closed proprietary architectures. On the other hand, this openness and the widespread knowledge of how these systems work have increased the knowledge base of those who would attack these systems. The fact that many of these open architectures are just beginning to build in or retrofit security services and mechanisms compounds the problem for these systems.

Distributed Security

Distributed security is inherently more difficult. To be effective, it must implement an organization's security policy in a consistent, comprehensive, and layered manner. It must be capable of keeping up with evolving network architectures and new protocols and services. Managing the configuration of a network is difficult enough. Making sure that security is preserved as changes are made during corporate reengineering and mergers is even more challenging. This is not aided by the plethora of point solutions for network security, which are dealt with later in this chapter. Network vendors' security frameworks merely position individual point solutions, which are not closely integrated together either functionally or through network management platforms.

Security Is Not Cheap

Strong security technology seems to be complex and expensive—not necessarily expensive in capital equipment terms, but rather in the complexity of management and the head count needed to care for it. The possibility that security management errors might also unintentionally block important information flows in an organization, or might somehow degrade overall network performance, are expenses most organizations would rather avoid. Much of the focus in security now is on how to obtain the strongest security with the most transparency and ease of use and the least impact on performance and operations.

What Is Network Security?

Security is really nothing more complicated than systems operating as they should. Regardless of whether "attacks" are intentional or unintentional, the network should continue to operate as the owner or subscriber intends. Unfortunately, specifying exactly how this should happen can be a complex exercise for any organization.

Policies for Network Security

These operational intentions are spelled out in the network security policy. There are three fundamental aspects to the policy, which can be thought of as the answer to these three questions:

- How do we control access to networked information systems?
- How do we control access to the intelligent devices that make up the network infrastructure?
- How do we safeguard the transport of information across the network?

It is the responsibility of the enterprise and its management to know what security policy is right for their organization. Deciding on a security policy and making it successful involves the entire organization. The policy must ultimately be based on operational business goals and must be closely aligned with revenue and market-share growth objectives. Protecting information assets is important, but making sure that the security policy is tightly coupled to the overall information technology strategy of the enterprise is crucial.

Security policies, when implemented seriously, require significant investments in time and capital. It is imperative that the policy be based on an honest appraisal of the total business environment and the consequences of not having adequate security. This appraisal should include an objective review of the internal and external threats, system vulnerabilities, and a risk assessment. Once these initial steps are complete, management is in a position to make the business decisions that result in an enterprise security policy. This policy statement then can drive the implementation of programs and procedures, standards of performance, measurable goals, and a methodology to monitor and adjust the overall security program.

With a security policy in place, the enterprise can begin identifying what security services must be present in an information system and where network security plays a role. Unfortunately, VLANs— particularly policy-based VLANs, despite their name—do little to ease the security burden, but can actually add to it.

Some Policies Are Not Technical

While this chapter discusses the technology of network security, it is important to remember that many powerful security policies rely on enforcing mundane practices, such as providing locked wiring closets in every facility to restrict physical access to network equipment and cabling.

Human resource policies are another important nontechnical area. Do employees who have given notice to leave retain all access to the network? How are passwords distributed to new employees? Is there an appropriate culture of security within the organization or certain critical departments? Does the employee handbook describe vacation policies in more detail than how to treat corporate information in an era of faxes, email, and mobile employees?

These areas of security are as important as many of the technical aspects of network security. Indeed, it can be tempting for people in technical positions to think of the problem primarily as a technical one and neglect the human element, which is the most difficult component of any policy problem.

Deciding Which Policies Need Network Technology

Network technology, while it cannot address all of an organization's security policy issues, must clearly play a role in solving many of the security challenges it has introduced. Deciding on what that role is and how it should be performed is the job of a network security architect.

A network security architecture specifies the required security services the network will deliver in support of an information security policy, and further specifies which hardware or software elements of the network will implement these services. When the network security architecture is completely defined, it explains how and where a network will answer the access and transport protection questions posed previously.

Not all security services required by an organization's information security policy must be supplied by the network itself. Some of the services can be provided by the information resources attached to the network. For example, a database management system (DBMS) can provide access control services to certain parts of the database, but the initial authentication and access authorization might be provided by a network registry or directory service that provides a single login capability. Or, in a situation where the network services are supplied by an outside service provider, the security policy might specifically require that all security management control be retained by the organization, and that critical security services be supplied by the customer premises equipment and software.

Policy for Network Security

Policy is a key aspect of network management, and thus of virtual LANs. Network security falls squarely in this policy area. And policy is the core of any security framework. In certain aspects, VLANs improve the basic security policy of a network. However, more should be expected of policy-based VLANs in terms of security.

Network Security Architecture

A network security architecture is a framework which organizes network security around six major security services: authentication, access control, confidentiality, integrity, nonrepudiation, and audit. These are the service categories essential to understanding network security. Table 7-1 summarizes each of these six services.

These basic services can be implemented in various ways using a number of different security mechanisms. For example, simple passwords are one form of authentication, while sophisticated time-based tokens, such as Security Dynamics' SecurID card, provide a more secure way to use passwords for authentication.

The exact mechanisms used depend on the particular network device and the communication services to be protected. These security services

TABLE 7-1

Security Services and Functions

Security service	Key function
Authentication	Confirms identity
Access control	Prevents unauthorized use
Confidentiality	Protects against unauthorized disclosure
Integrity	Detects unauthorized modifications
Nonrepudiation	Provides proof of origin and delivery
Audit	Provides records for independent review

constitute the building blocks that you can use to build the network security architecture appropriate to your organization's network architecture and security policies. In addition to providing an architecture for security services and individual security functions, a secure network architecture provides the basis for building secure, reliable, and highly available enterprise networks. Security is a multifaceted puzzle, not a one-dimensional problem.

Security Services

Networks can play a wide variety of security roles. Sometimes the security services can be almost transparent; at other times they are very visible such as when logging into a server. The security services the application users of the network expect may be very different from those needed by the operators of the network.

Many security services are probably already familiar. They represent different aspects of security and enable a user to more finely tailor network security for consistency with the organization's security needs.

The conventional Open Systems Interconnection (OSI) reference model defines a set of communications services that are distributed over the well-known seven-protocol layer stack. Part 2 of that model (ISO 7498-2: Security Architecture) performs a similar function for security, formally defining security services and making recommendations on which services are appropriate to which layers. Just as the OSI reference model has its limitations, the OSI security architecture is similarly restricted. Although useful as a conceptual framework defining concepts and their applicability, as an implementation guide it is much too limiting (see Fig. 7-2).

The six security services are discussed, followed by a discussion of the various mechanisms that can be used now and in the future to implement and strengthen a security policy that meets the business needs of the enterprise.

Authentication

The data authentication service provides the user with confirmation of the identity of the remote communication entity or that the origin of the received data is that remote entity. This service increases the user's confidence that the remote entity is not masquerading as, or impersonating, a different entity or is replaying data from a previously authentic

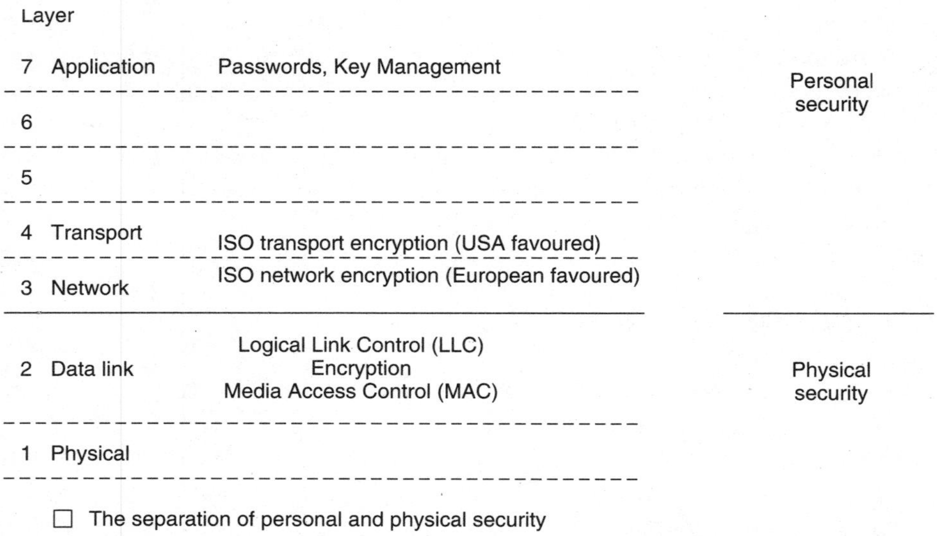

Layer

7	Application	Passwords, Key Management	Personal security
6			
5			
4	Transport	ISO transport encryption (USA favoured)	
3	Network	ISO network encryption (European favoured)	
2	Data link	Logical Link Control (LLC) Encryption Media Access Control (MAC)	Physical security
1	Physical		

☐ The separation of personal and physical security

Figure 7-2 Network security standards and the OSI model

connection. For many security policies, authentically establishing the identity of a user, server, or communication device is a critical step on which all subsequent security actions will be based.

Access Control

The access control service prevents unauthorized use of a resource, which includes precluding the use of an authorized resource in an unauthorized manner. Once a user (communication process, application, or actual human being) has been identified, usually by the authentication service, access control can determine what resources should be made available. This service can be used to control the type of access (e.g., read, write, execute) and can be used to protect network resources or external resources that are attached to the network.

Confidentiality

Confidentiality service protects data from unauthorized disclosure. In many ways, it is a variant of access control, but the emphasis of this service is preserving the privacy of communications. Confidentiality can be applied to all data carried in a communications protocol, or selectively to particularly sensitive fields within a protocol (e.g., a credit card number). In the most extreme instance, traffic flow confidentiality can be used to hide the identities of the communicating parties, the volume of their traffic, or the very fact they are communicating at all.

Integrity

The data integrity service is directed at active threats and detects modifications of data as it passes from one user to another. This can be changes, insertions, deletions, sequence changes, replays, or reflection of data exchanged between two users. Integrity is closely related to authentication, and, in many ways, it is very difficult to achieve one without the other. Knowing that data has come from a particular user (authentication) is not very useful, if it is possible that the data has been modified or is a replay of a previous communication (integrity).

Nonrepudiation

The nonrepudiation service is possibly less familiar but provides proof as to the origin or delivery of data. The key word here is *proof;* that is, evidence is created by the service that can be provided to a third party demonstrating that the communication must have come from a particular party or that it was, in fact, delivered to the intended destination. Thus, this service is particularly applicable to commercial transactions where both the sender or receiver must be protected against the false denials of the other. Registered mail, return receipts, and notarized and witnessed contracts are examples of some common methods society has developed to provide this service in everyday life.

Audit

The audit service provides records to support an independent review and examination of the network to test the adequacy of security controls. Also implicitly included with the audit service is the means for detecting events that are security relevant and the use of other services to protect the audit records. ISO considers audit trails and event detection "pervasive security mechanisms." The ability to securely report, log, and initiate recovery actions is crucial to operating a secure network and warrants its own service focus. Corrective feedback is an essential part of operating any quality system. The audit service is the key to determining how well the security architecture is meeting its security quality goals.

Security Service Mechanisms

If security services establish the basic attributes of security that we are striving to achieve, then security mechanisms are the actual methods and techniques that we devise to accomplish them. In general, the simpler the mechanism, the greater the possibility that mechanism might be defeated and the service rendered ineffectual. Thus, as a security policy evolves, it must necessarily establish minimum standards of security mechanism strength. Users who seek "more security" for their network are frequently, in effect, asking for stronger security mechanisms to support the security services they have. In other words, the service may be perfectly correct for the organization's security policy; it just has not been implemented with mechanisms that are strong enough to thwart the threats being mounted against it (see Fig. 7-3).

Some security mechanisms are very familiar, but perhaps have been used to support other network services besides security. A common example is packet filtering. Simple packet filtering has been used to tune performance or to confine certain classes of traffic within an organization's internetwork, or as a troubleshooting technique. Recently, more elaborate filtering techniques have been developed as part of a firewall access control strategy or to automatically route traffic over more protected paths.

The list of mechanisms that can be used in supporting security services is long, but how these mechanisms are used in the different product areas can differ. For instance, a password mechanism is universally used for authentication. Other mechanisms, such as encryption, can be used to support almost all services.

Authentication Mechanisms

Authentication is an essential service. Most commercial organizations, consciously or not, operate under identity-based security policies. Net-

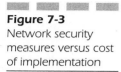

Figure 7-3
Network security
measures versus cost
of implementation

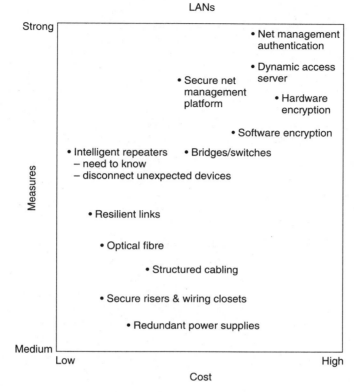

work privileges and access are authorized on the basis of the user's identity. Securely and reliably confirming a user's identity is, therefore, fundamental.

One note of caution. It is sometimes easy to confuse who or what is being authenticated. In some cases it is a living individual. It also might be a computer or program representing that individual. This will be referred to as *user authentication*. In other instances the entity being identified is strictly a network device that may be associated with many "users." In that case, this will be referred to as *device authentication*.

Passwords. *Passwords* or *personal identification numbers* (PINs) are the most commonly used authentication technique, representing something a user knows and secretly shares with the device at the other end. Passwords are typically submitted as part of a two-way handshake that should be part of logging in to hosts and servers, remote access servers, and the management ports of critical devices such as routers. Passwords can also be used within management applications to identify those who have been granted additional privileges for particularly sensitive functions (e.g., managing sensitive monitoring devices such as RMON probes). Passwordlike values can also be used in protocol messages exchanged between devices. SNMP community strings, RIPv2 authenticators, and PPP (Point-to-Point Protocol) PAP (Password Authentication Protocol) passwords are examples. Passwords can be exchanged in both directions to obtain mutual authentication. Passwords are subject to protocol analyzer attacks and unauthorized reuse. The use of authentication servers to act as centrally administered (and protected) password databases is also recommended where possible. Protocols such as RADIUS can be used to securely transport passwords between the access device and the authentication server.

Challenge-Response Handshakes.   The *challenge-response handshake* mechanism achieves stronger authentication by using a method in which one side initiates an authentication exchange, is presented with a unique and unpredictable challenge value, and, on the basis of a secretly shared value, is able to calculate the appropriate response and send it back. This three-way handshake provides a "liveness" check and defeats simple password sniffing attacks. PPP CHAP (Challenge Handshake Authentication Protocol) is an example in which the keyed-MD5 (Message Digest 5) authentication algorithm is used to calculate the proper response. CHAP can be used to provide user authentication (e.g., client to remote-access server) or device authentication (e.g., router to router).

The use of a centralized authentication server (e.g., RADIUS server) can be used to limit the distribution of the shared secret to just the server and the users or devices requesting the authentication service.

Callback.   *Callback* can be considered another form of handshake, commonly used in dialing in over the telephone to remote-access servers. These servers can be configured to dial back a specified number associated with the user. While this has limited use for the mobile user, it can be useful in enforcing authentication if a password is lost or "loaned." It is also useful for audit and billing purposes and can, in some cases, reduce telephone costs.

One-Time Passwords.   The *one-time password* mechanism uses a technique in which the password provided is never reused. This defeats protocol analyzer attacks because any password intercepted can no longer be used again. One of the most common examples of this technique uses time as the constantly changing value on which to base the password. The popular SecurID card displays a value on its LCD (liquid crystal display) screen that changes every minute. This value (something the user has) plus the user's PIN (something the user knows) can be submitted to the authentication server, where it can be uniquely compared against the value computed for that user's card at that particular time. One-time passwords are the next logical progression for protecting remotely accessed devices. They are particularly important in cases where logins are initiated over the network via remote Telnet. A variation of this technique, using a set of encrypted counters, can be used to authenticate SNMP SET commands from a network management console.

Kerberos Authentication.   *Kerberos* is a more complex method for authenticating network users and protecting network traffic based on DES encryption. As an Internet Engineering Task Force (IETF) standard, it is gaining a strong following, primarily in the UNIX community.

Kerberos works by having a central server grant a "ticket" honored by all networked nodes running Kerberos. Tickets are encrypted, so passwords never go over the network in clear text. The single login feature of Kerberos means that users do not need to reenter their password when accessing a different computer—they just submit a ticket. The granting of a ticket itself can be used to authenticate the user logging into a remote access server. In a full Kerberos deployment, the ticket can be retained by the client and submitted as needed to gain access to other resources without having to log in again.

Finally, Kerberos tickets can be used to establish full confidentiality as well (see the section on Kerberos encrypted login). When operating in this mode, the encryption can be used to protect the communication flow continuously, and thus preserve the authentication of the connection beyond the initial setup.

RADIUS.   The *Remote Authentication Dial-in User Service* (RADIUS) protocol is rapidly becoming the basis for future, open-standards-based exchange for all types of authentication data (e.g., PAP passwords, challenges and responses). Using RADIUS, a client can exchange authentication, access control, accounting, and device configuration information with a RADIUS server. The RADIUS server can authenticate the user or device using its database of user IDs and authentication parameters. The keyed-MD5 algorithm provides authentication and integrity for the RADIUS messages and selective confidentiality for authentication parameters they might contain.

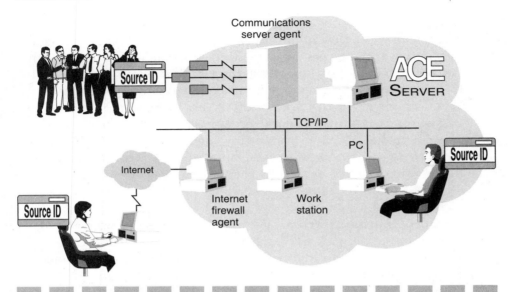

Figure 7-4 Security Dynamics' SecurID software implementations (Source: Security Dynamics)

Further discussion of RADIUS is contained in the sections on access control and auditing.

Security Tokens. *Security tokens* are small portable devices that provide secure storage of authentication parameters and the processing capability to participate in one-time password or challenge response authentication exchanges. In addition to the SecurID card from Security Dynamics, other popular tokens include those from Enigma Logic and Crypto Card. These tokens and their associated servers can form the basis for an authentication method that covers client, server, and host security as well as communication devices such as remote access devices (see Fig. 7-4).

Access Control Mechanisms

The fundamental purpose of all access control mechanisms is to prevent unauthorized use of a network resource.

Access Control Lists. The *access control list* (ACL) is at the heart of most access control mechanisms and constitutes the basis for making authorization decisions. The list can be either inclusive (specifying those users who are authorized—all others are unauthorized) or exclusive (specifying those users who are not authorized—all others are authorized). Lists can be organized by user, user group, device, or device port. User group profiles can be used to simplify the administration of lists with many users. Lists can be centrally administered off line and distributed to the various devices or maintained on line in a central set of servers. Devices needing to determine access privileges refer to their local lists or communicate (securely) with the central server to obtain this information. Most networks rely on a mixture of both approaches for both security and reliability reasons.

A common network device example is a list of IP nodes authorized to access embedded services such as SNMP, Telnet, FTP, or TFTP. SNMP agents associated with critical devices such as routers or RMON probes are especially in need of such protection since the community string "passwords" are too easily learned.

Finally, because many networks use the access control lists built into their host and network operating systems, it is important that there be a common method for administering the lists maintained in network devices and network operating systems. Interfaces between network access control mechanisms and UNIX, NetWare, and Windows NT mechanisms will become increasingly important to ensure manageable and consistent access control lists.

ACL Violation Logging. Providing strong firewalling capabilities is not enough in today's world of networking; network managers need an option for centralized reporting. In the past, network managers did not know that they had been hit by a hacker until the damage was done. At that time, the only available early warning tool was to scan host log files. Although this is still an excellent method for security diagnostics, it does not scale well. ACL reporting tools help managers by providing violation information and prevention at the network perimeter.

Authorization Servers. RADIUS servers and other authentication servers are examples of a centrally administered approach to access control in which the server contains not only a means for authentication but also a database containing the services that the authenticated user or device is to be granted. In the case of RADIUS, the server can download configuration parameters that can enable authorized communications and security services. RADIUS-supported security parameters include user-specific filters and authorized protocols.

Communication services examples include session control, channel reservation, and channel aggregation across multiple lines. To ensure scalability, these servers should also support proxy services that allow one server to receive access control information from another server responsible for a particular user.

Traffic Filtering.   One of the most basic and effective methods for enforcing access controls is at the data-packet layer. In its simplest form, a filter can block packets that do not conform with security policy rules and to forward those that do. Filters can be assigned to each port and traffic direction (i.e., incoming or outgoing) and be based on source and destination address, protocol type, and other fields within the packet. Routers can be particularly effective filtering engines because they are optimized to make rapid forwarding decisions based on these very same attributes. Filters can be used in switches and remote-access servers as well. Besides blocking or forwarding, a filter should also be able to trigger an event for logging and/or alarming.

Policy Filters.   Formerly called *route filters, policy filters* are available on routers and enable the network manager to set up an access control policy specifically aimed at controlling access to and from the routing tables. "Accept" filters are used to specify which route updates will be accepted from which routers or networks.

Intrusion Detection.   One method, similar to filtering, available to control access to LANs is based on the ability of hub repeaters and switches to capture and compare the source address of all packets received from the connected nodes. This capability, which operates at full wire speed, can be used in conjunction with "allowed nodes" access control lists, to ensure that only authorized stations are connected to the LAN at the appropriate physical location. On the basis of security policy, detection of unauthorized nodes can result in a security event, alarm, termination of the offending level 2 MAC address (if multiple workstations are attached to a port), or shutting down the port itself.

For ISDN networks, which supply incoming *calling-line ID* (CLI) screening, CLI inclusive screening can be used to accept or reject incoming connections after comparison with an access control list to authorized CLIs. Note that if the ISDN call traverses the networks of two or more interconnected carriers or a telecommunications switch lacking CLI features, then the calling-line ID will not be presented to the access control server for screening. Instead, a message such as "CLI unavailable" is transmitted. The same problem also occurs with remote dial-in.

Inactivity Timeout.   Remote access servers can be configured to automatically terminate user sessions when there has been no activity over the line for a specified period of time. This reduces unnecessary security exposures from unattended terminals and also saves on phone charges. A similar facility can be used with Telnet and other connection-oriented network protocols.

Routing.   Routers, in conjunction with filters, can also play a role in determining which network resources will be granted to a particular stream of packets. For example, IP delivery options allow a network manager to specify in advance the most appropriate path for certain classes of IP traffic with respect to resource considerations. With this capability, traffic arriving from certain insecure networks might be routed via networks having more security or less valuable attached resources. Access to paths that traverse subnetworks that serve particularly sensitive applications would be implicitly denied.

Security Labels.   The *Revised IP Security Option* (RIPSO) (RFC1108) specifies an optional IP header field that contains a security classification and handling label. Used primarily by the U.S. government, this label allows the hosts and routers to appropriately label IP traffic on the basis of its security level. Other hosts and routers can then check the labels of arriving datagrams. According to the rules specified in RFC1108, traffic may be accepted, rejected, or forwarded on the basis of the security attributes of the host or the attached networks. Although not used by most commercial organizations who use identity-based security policies, labeling is essential to government mandatory access control polices based on security clearances and information classification levels (e.g., "top secret").

VLANs as Security Domains

The capability of current VLANs to create firewalls can also provide more stringent security, replacing a lot of the functionality of a router. This holds true when VLANs are implemented in conjunction with private port (single-user) switching. The only broadcast traffic on a single-user segment would be from that user's VLAN (i.e., traffic intended for that user). Conversely, it will be impossible to listen to broadcast or unicast traffic not intended for that user—even by putting the workstation's network adapter in promiscuous mode—because the traffic does not physically traverse that segment. There are exceptions to this, which cause security as well as management problems which will arise in future VLAN implementations. These are highlighted later in this chapter.

VLAN Tagging

The virtual LAN capabilities of switches allow the network administrator to establish virtual LAN groups that directly parallel an organization's security policy. For example, finance, engineering, and marketing VLANs can be established to ensure that tagged information flows can be logically segregated. VLANs dedicated to network management uses are another possibility. Controlled communication between VLANs can be implemented using any of the other mechanisms discussed above— not just the screening router as most vendors' PowerPoint slides suggest.

Access Control for VLANs

Switches are the key to providing access to VLANs, since intelligent switches with traffic filters can restrict traffic to groups of ports, groups of MAC addresses, and/or groups by protocol. The same mechanism which provides basis of the virtual workgroup also provides the basis for VLAN access control security. A further security check can be applied to the MAC/VLAN membership ID address tables by checking that they have not grown unexpectedly in size through a malicious user masquerading as an authorized member of a VLAN. In addition, the size of address tables limits the maximum number of attached users, which may be useful in determining which of the membership has malicious intent.

ATM LANE Client Registration

LANE services allow administrators to establish a number of security relevant access control policies. For example, selected clients can be preassigned to a specific Emulated LAN. In this case, all client registration requests are ignored and the client is placed in a specified Emulated

LAN or a special default Emulated LAN. The default Emulated LAN can be further restricted to prevent multicast or broadcast frame transmissions, thus effectively preventing communications between these clients via data-direct SVCs.

Protocol Gateways

Gateways which translate from one protocol suite to another can perform access control features as well. For example, a type of firewall can be established between a local NetWare LAN running IPX and the Internet running IP. IPX workstations running IP can access the Internet without any IP traffic traveling between the Internet and the local IP LAN. IP traffic from the Internet is blocked by this gateway as well as effectively hiding the local IP addresses and configurations.

Firewalls

Routers have had a "firewall" function since they were first invented. Originally, a router served as a barrier that could confine protocol storms to one subnetwork. Subsequently, simple traffic filters could be used to prevent certain types of services [e.g., File Transfer Protocol (FTP)] from being allowed between networks.

As users' needs for these services increased, simple all-or-nothing filtering became insufficient and special-purpose security gateways, which did little or no routing, were developed to concentrate on enforcing security rules on the traffic flow. Today firewalls offer a range of services: static traffic filters, dynamic (or stateful) filtering, circuitry-level gateways, and application proxies. In almost every firewall deployment, screening routers are used to perform static filtering to create a "demilitarized zone" (DMZ), which provides a first layer of protection to the firewall and the inner network.

The distinction between static filtering done in routers and dynamic filtering and proxying done in firewalls is vanishing—for example, Cisco routers will soon incorporate Checkpoint Firewall-1; Bay Networks and 3Com routers already do. Common management of the access rule bases between the two devices will continue to evolve (e.g., Checkpoint Firewall-1 Management Center), followed by a merging of some functions into a single router-firewall platform. There is no reason why this trend should not continue and both static and dynamic filtering will eventually become part of layer 3 switching and thus VLAN implementations. In fact, the first example of a firewall implementation in a layer 3 switch is Alcatel Data Networks' Alcatel 1100 LSS, which implements an integrated security firewall based on virtual LAN connections. One password enables users to gain access to multiple VLANs within a given network (based on the Alcatel kit, obviously). In addition, the network management of the IP firewall is integrated to some extent with ADN's SNMP-based network management system. Alcatel Data Networks may be the first, but many will follow its lead.

Confidentiality Mechanisms

Confidentiality mechanisms protect communications from unauthorized disclosure. While encryption is the most flexible tool that applies here, a number of basic techniques are also discussed because they can often be deployed quickly and with less system impact.

There is no single "best" place to perform encryption, and as applications evolve, encryption will become more commonplace in both the application and the supporting communication protocols.

Eavesdrop Protection

Operating at full Ethernet speeds, eavesdrop protection hardware in Ethernet repeaters and hubs can immediately determine the destination MAC address of the packet and compare it with the authorized addresses for each port in the hub. According to the MAC address, the packet data can be overwritten or erased for all unauthorized ports. With this totally transparent capability, the risks are greatly reduced that sensitive traffic and user passwords can be intercepted by anyone plugging a PC into an empty office outlet or Trojan horse protocol analyzer programs introduced by an outside hacker with UNIX root privileges.

Access Control Filters, Firewalls, Labels, and Tags

Traffic filters, firewalls, security labels, and VLAN tagging can be effective in blocking information that should not be allowed to leave protected networks or enter unprotected networks or network segments. These access control mechanisms can be used to enforce security policies that have identified particular networks, hosts, or access points as particularly vulnerable to unauthorized disclosure.

Routing

Routers can also play a role in determining secure paths for sensitive data. For example, in conjunction with filtering, IP delivery options allow a network manager to specify in advance the most appropriate path for certain classes of IP traffic with respect to confidentiality considerations. Similar capabilities exist in an ATM context and can support explicit control over how call setup requests are routed and more importantly, which calls are routed at all. If a portion of the network is deemed unsecure, it is possible to specify that no route is acceptable and the call request is refused.

Policy Filters

Policy filters are available on routers and switches with routing capabilities. They enable the network administrator to set up policies specifically targeted at controlling the dissemination of route updates. Announce filters specify which routes will be announced to other routers, thus protecting sensitive network route configuration information from external networks and intruders, which have no need to know this information.

Encryption

Link Encryption. The ability to encrypt the payload of a packet at the data-link layer is essential for many applications, especially for private or public WANs. Routers which sit at the LAN/WAN boundary are particularly appropriate locations for this capability. Depending on bandwidth requirements, encryption can be done in either hardware or software.

Software has the advantage of being more portable across different communication platforms if the processing power is available. The ability to remotely manage the keys at each end of the link is essential. Of equal importance is the capability to automatically change the traffic keys in use, frequently, transparently, and without operator intervention. Hardware-based encryption, especially if it integrates into the modular and scalable architecture of today's router platforms, can provide high-performance encryption processing for the full range of LAN and WAN interfaces.

Network Encryption. To extend confidentiality across the extended enterprise, it is imperative that a fully interoperable method of encrypting IP datagrams be available. For this reason, the IETF's IP Security (IPsec) Working Group has produced the IP Security Protocol (IPSP) suite. RFC1825 to RFC1829 specify an IP-level security architecture, encryption [ESP (Encapsulating Security Protocol)] and authentication (AH) protocols and associated data transforms. These RFCs are currently proposed standards and have already formed the basis for such interoperability initiatives such as the Secure WAN (S/WAN) consortium sponsored by RSA Data (now part of Security Dynamics).

Secure Tunneling. IPSP encryption (ESP) can operate in both transport and tunnel modes. In transport mode, the datagram payload is encrypted, and the original IP header is left in the clear. In tunnel mode, the entire original IP datagram is encapsulated, encrypted, and placed in the payload of the new datagram. This has the advantages of hiding the original user addresses and the flexibility that allows intermediate nodes (e.g., routers, firewalls) to act as proxy destinations for traffic targeted at a particular end system. The net result is that encrypted tunnels can be established between gateways that securely connect enterprise LANs across insecure networks. The encrypted tunnel, depending on the application, can serve as a cryptographic conduit to prevent unauthorized access to the traffic (i.e., confidentiality).

Conversely, the tunnel can act as a secure pipe to confine user traffic to authorized paths within the network (access control). Consequently, managing these tunnels' endpoints is just as important as establishing the security between the endpoints. Static configuration, dynamic discovery, redundancy, and automatic switchover are essential to ensure that encryption remains transparent and does not degrade any of the dynamic and redundant features that contribute to the reliability of routed networks. Tunnels can be extended directly to workstations and servers if needed.

Key Management. Further work is under way within the IPsec

Working Group to expand the IPSP suite to include an Internet Key Management Protocol (IKMP). The most promising Internet drafts include Simple Key Management for Internet Protocols (SKIP) and the Internet Security Association Key Management Protocol (ISAKMP) using the Oakley key determination protocol. Both SKIP and ISAKMP/Oakley require the authenticated exchange of public key certificates to be most useful. The Public Key Infrastructure (PKIX) Working Group is standardizing the format of the Public Key Certificates and the Certificate Authority requirements needed for the successful operation of these protocols. Plans are under way within the S/WAN initiative to conduct interoperability testing using one of these proposed IKMPs.

Encryption Interoperability.   The S/WAN initiative is seeking to ensure full interoperability of encryption between routers, firewalls, and other intermediate packet encryption devices, as well as hardware and software implementations of IPSP in the end nodes. This interoperability must be at all levels: algorithms, data transforms, protocol data unit formats and processing, management of keys, negotiation of security parameters, and establishing security associations. The chain of trust, on which organizational and interorganizational security associations will be based, will have public key certificates as its basic medium of exchange.

Encryption Performance and Scalability.   Data encryption is similar to data compression in that it is computationally intensive. While software encryption is certainly feasible, performance in a typical server or router environment will probably be limited to around T1 speeds for generally accepted algorithms such as DES. Security that limits the performance of a network is not acceptable in most environments. For this reason, as well as the advantages of implementing the algorithms and key storage in more tamper-resistant hardware, encryption should be available in high-performance hardware for high-speed backbones such as ATM. However, encryption devices which operate at ATM speeds—such as StorageTek Atlas—are very expensive and thin on the ground. These cryptographic accelerators should be modular so that one or more can be configured to provide the security horsepower needed to ensure transparency on backbones as well as remote links.

Encryption and Troubleshooting.   The widespread use of Data Encryption Services can have an unintended impact on today's methods of monitoring and troubleshooting networks using network protocol analyzers and RMON probes. With the use of link or IP layer encryption, the upper-level protocol control information is not visible to conventional protocol analyzers, and even true source and destination address information can be hidden when operating in secure tunnel mode. Cryptographic management must include features for temporarily disabling encryption for troubleshooting purposes or preferably providing selective decryption for monitoring tools. Access to these features must, of course, be securely controlled.

Encryption and Data Compression.   Encryption, if not implemented correctly, can also impact the performance of data compression. It is important that any data compression be performed first, prior to encryption. Encrypted data is essentially randomized, without the redundancy of clear text, normally exploited by the compression algorithms. This means that IP encryption followed by data-link compression will result in a degradation in overall throughput when compared to compression without encryption. In systems where compression is an important requirement, integrated compression-encryption processing must be carefully architected to ensure that the benefits of compression are not lost.

External Encryption Interfaces.   In some environments, it may be appropriate to use standalone encryption devices that interface to network devices such as routers. Traditionally, U.S. Department of Defense applications require this approach. Network devices should continue to support the direct connection of new government and nongovernment encryption devices that comply with standard network interface, management, and protocol standards.

Kerberos-Encrypted Login.   Stronger authentication and continuous protection of the login session can be provided by the use of Kerberos. Kerberos clients can be granted tickets by the Kerberos security server that contain the keys for fully encrypted sessions with "Kerberized" servers. Encryption between client and server provides very strong protection against session hijacking. For example, if Kerberos is used only for initial Telnet authentication, TCP sequence number attacks can be launched to gain control of the authorized session.

On the other hand, if the Telnet session is fully encrypted, the attacker must overcome both the encryption and the TCP sequence checking mechanisms—a much more daunting task. Encrypted login can be particularly important when accessing the maintenance and configuration features in critical network devices over the network.

Directory Services. Two authentication encryption techniques are commonly used in directory services: symmetric and asymmetric.

Symmetric cryptography (also known as *private key*) uses a method where sender and receiver both share the same encryption-deencryption algorithm. Private-key encryption is fast and efficient but has a significant drawback. Both the sending server and the receiving server store and keep track of the key for each item. This can be difficult to do safely. This problem led to the development of public and private key encryption. Asymmetric cryptography (also known as *public-private key*) uses two different keys that have a mathematical relationship to each other:

- *Public key.* This key is assigned to an object and can be published openly to anyone wanting to send a message to that object.
- *Private key.* This key is assigned to the same object, which keeps it secret.

In public- and private-key encryption, either key can be used for encryption and the other is then used for deencryption. Because the mathematical relationship between the keys is difficult to compute, the public key can be made available freely without the risk of exposing the private key. If you have the public key, you can encrypt messages to a client that possesses the matching private key. You cannot use the public key to deencrypt messages that have been encrypted with the public key. Only the private key can deencrypt the messages encrypted with the public key. Because only one object has the private key, only that object can deencrypt your messages. Conversely, a sender can encrypt data using the private key. The recipients of this message use the public key to deencrypt the message. If the deencryption is successful, the recipient can be sure that the message was encrypted with the corresponding private key. In this case, many objects can deencrypt the message, but only the holder of the private key could have generated the message.

Integrity Mechanisms

Integrity mechanisms provide a way to detect unauthorized modifications of the network communication. In many ways, integrity is one of the most important services. Applications and networks cannot run correctly if the data they are using has been tampered with. Detection of tampering by an attacker who is fully knowledgeable of the applications and protocols requires cryptographic techniques similar to those used for authentication.

Checksums. A *checksum* is a mathematical calculation, producing a result that depends on the entire content of the message. This redundant piece of information can be inserted in the message by the sender and recommitted by the receiver to see if the message has been changed. In the case of communication environments, the check is done to ensure that no errors in transmission and reception have occurred because of noise problems or processing errors. Mathematical algorithms, variously known as *cyclic redundancy checks* (CRCs) or *frame-check sequences* (FCSs), have been developed to detect errors with high probability. A knowledgeable attacker can, of course, make changes to the message and the checksum to prevent detection. Encrypting the message and checksum can defeat simple attacks but typically are not sufficient against replay or cut-and-paste attacks.

Message Authenticator. Secure, one-way hash functions efficiently reduce an arbitrary message string into a fixed-length quantity (hash). Knowing the hash alone, it is difficult to independently reconstruct the original message. And, unlike checksums, even knowing the original message, it is difficult to compute another message that would have the same hash. If a shared secret key is used as a piece of the message but is not actually sent, the resulting digest can be used to detect deliberate attempts at modification as well as serve as an authenticator of the message's origin. This method also has the advantage that the message contents are in the clear and can be used by any recipient. However, only those possessing the shared secret can verify the integrity of the message. (If only the sender and receiver share the key, then origin authentication can also be obtained). MD5 and SHA-1 are commonly used hash algorithms. MD5, for instance, has been specified, router-to-router protocols (RIPv2 and OSPF), IP security (IPSP-AH), and RADIUS communications. This same mechanism can be used to protect file transfers, including user data and the configuration and software images downloaded to communications devices.

Sequence Numbers. Sequence numbering can be used to detect message playbacks and deletions. Message authenticators or encryption can be used to protect the sequence numbers included in connection-oriented protocols (e.g., TCP) or can be specified within a security protocol to provide this service.

Nonrepudiation Mechanisms

These mechanisms provide proof of the origin and/or delivery of the communication. Generally, this has been considered an application-oriented service. Consequently, various management applications are the most obvious place to look for this service.

Digital Signature.   Public-key encryption systems, in which the encryption key is different from the decryption key, has made digital signatures possible. One key is kept secret by the originator and used to encrypt or "sign" messages. The other key is widely distributed to all potential recipients and is used to decrypt signed messages to "validate" their origin. The original message, the "signed" version, and the public validation key all constitute the proof of origin required for nonrepudiation. Signed return receipts can also be used as proof of delivery. The most common signature algorithms are RSA and the Digital Signature Standard (DSS).

Public-key-based digital signatures have many applications, but in the network management space they will be adopted initially in key management protocols as part of establishing a security association (e.g., ISAKMP/Oakley). Individual devices and users will have public keys that will be exchanged as part of that protocol. These public keys will be contained in digitally signed certificates issued by trusted certificate authorities.

These authorities can be internal to an organization, but more likely will be third parties that are trusted by many different organizations to securely issue and sign certificates and manage renewals and revocations. This so-called public-key infrastructure, pioneered in the X.509 Directory Services standard, will be the basis for building a secure communications and application infrastructure. The IETF is working on the necessary standards in the IPsec and PKIX Working Groups.

Audit Mechanisms

Audit mechanisms provide independent records of communications activity that enable managers to measure the effectiveness of the security services provided as well as detect any suspicious activities. These records form an objective basis for making changes to the security services and investing in stronger mechanisms. They also form the basis for adjusting security policies, conducting investigations, and, if properly protected and handled, used as evidence in prosecutions. The value and importance of these records is attested to by the fact that they are among the first things modified by sophisticated attackers after an initial break.

Audit Trail

Audit trails are the sum of all records collected and retained that relate to normal and abnormal events, occurring during the daily operation of a system. Network audit trails relate to the events that are noted by the set of devices making up the network. These records should be time-stamped and contain information identifying the system, the communicating entities (e.g., user IDs, addresses, ports), the particular event, and other event-specific data. Synchronized system clocks can make the correlation of date-timestamps easier during an audit trail analysis.

Typically, audit trails begin as recorded events in the network device. They can result in immediate notifications (e.g., SNMP traps) or can be stored in an internal event log. The manager should be able to select which events should be recorded. A wide set of criteria should be supported [e.g., protocol type, event severity, event identifier(s), subsystem/module] to identify what should be included or excluded from the log. The ability to capture and store files locally or upload them as part of the permanent trail is, of course, required. Event logs and the collected audit trails should be properly protected. Authentication, access control, integrity, and confidentiality mechanisms should be in place to ensure that unauthorized tampering with the event records and the recording criteria can be prevented, or at least detected.

Event Monitoring

Events can be monitored either centrally or on an individual device. Regardless, systems should be in place to ensure that the local records can be periodically gathered and protected in a centralized manner for regular analysis. Filtering criteria [e.g., protocol type, event severity, event identifier(s), subsystem/module] can again play a role to keep the volume of records forwarded for centralized monitoring to a manageable level. The unfiltered local log can be consulted for more information. Similar display filters are also desirable for quick viewing of records of interest. Syslog messaging is a very common method for transporting logged events to a UNIX system running the syslogd daemon. From there the logs can be dispatched to designated hosts, files, or printers.

Event Alarms

SNMP traps can be directed to the appropriate management station for real-time notification of specific events. These alarms should be logged and should be capable of initiating the usual visual, audible, and pager alerts, based on the severity of the alarm or warning.

Accounting Records

Although not strictly part of security, accounting services and mechanisms can be frequently incorporated into an overall audit trail picture of network activities. Accounting and billing mechanisms, such as those included in the RADIUS accounting protocols, are being rapidly standardized as part of dial-in user services. The use of accounting data (e.g., user name, calling number, type of service, session lengths, packets and/or bytes transferred) should be considered supplementary to event logs, alarms, and audit trails.

Product Categories

These services can be provided across several major product categories in the network infrastructure. For clarity and understanding, these can be separated into five basic product types:

- LAN access
- Router services
- Remote access
- Directory services
- Network management

LAN Access

The importance and popularity of remote networks pose issues for network managers who administer and impose security policies governing logins from remote sites. Current remote-access security solutions were designed for single-user, modem dial-up connections and do not ensure the integrity of data transmissions from remote networks that have multiple users and devices.

Today's distributed computing has also extended to a mobile world, where workers want access to the corporate site from ever-changing locations (as described in Chap. 2). Branch offices with WAN connectivity also want to provide access and authorization privileges that are unique to each end user. The home office today is changing as well, from a single-user using a dial-up connection such as a modem line or Integrated Services Digital Network (ISDN) service, to a multiple-device telecommuting environment. Many small offices and home offices (SoHos) have LANs, and their users want to connect all their resources to a central site. At the corporate site, the network manager must find a way to protect the main corporate computer from unauthorized access.

The shift from the centralized mainframe environment to the distributed client/server environment has forced network managers to move from a central security policy to one that allows for access from remote locations. While remote logging provides flexibility for users, it also poses security problems for the corporate or campus network manager.

Today's most common security solution is to use static access control lists—lists manually created by the network manager that define who can access the network—to authenticate and authorize remote users. In today's world, network activity provides the opportunity for break-ins by network hackers. An added vulnerability is that static access lists containing a wide range of network addresses can be stolen by the hacker.

Current security solutions exist, but many have drawbacks, including

- Firewalls or bastion hosts functioning as gatekeepers at the periphery of a network. This additional equipment is expensive.
- Authentication by means of static access lists. Standard access lists are stored in nonvolatile random-access memory (NVRAM), restrict who can access the system, and provide no challenge mechanism beyond a network address.
- Software solutions that require client software and applications to be modified to support them.

■ Application software-based security policies, such as using a specific network application to make connections and send electronic mail and files. There is no methodology to force users to choose a particular application for transmitting sensitive information.

Where multiple systems and applications are in use, security policies based on application layer security are often difficult to enforce. An overlapping security solution approach is prudent; any single security solution is more vulnerable.

Router Services

At the level of the individual network device, the security of router configuration (or the routing function) in VLANs is critical, since it operates as a traffic police officer, routing traffic between VLANs. This is because alterations in router configuration can significantly affect network connectivity and function. Telnet or Web browser connections to the router are standard methods managing a router, including its configuration and protocol setup. This can be improved through the network management system providing authentication services using the hard tokens (such as Security Dynamics' SecurID token cards). Operating in conjunction with an authentication server, anyone seeking Telnet or Web browser access to the router must know the password and possess a token corresponding to that password.

Remote Access

Remote-access servers provide communication facilities for traditional "dumb" terminals, terminal emulators, workstations, personal computers (PCs), and routers, using a serial line framing protocol such as Point-to-Point Protocol (PPP), Serial Line Internet Protocol (SLIP), Compressed SLIP (CSLIP), or AppleTalk Remote Access Protocol (ARAP). In other words, a network access server (NAS) provides connections to a single user, to a network or subnetwork, and to interconnected networks. The entities that use the NAS to connect to a network are called *network access clients* (NACs); for example, a PC running PPP over a voice-grade circuit is a NAC (see Fig. 7-5).

The authentication facility provides the ability to authenticate a user beyond a login and password dialog. A user can be challenged by a

Figure 7-5
Remote-access security requires a three-tiered approach. This identifies the user, protects transmitted data, and protects valuable network resources. Each type of protection requires different techniques

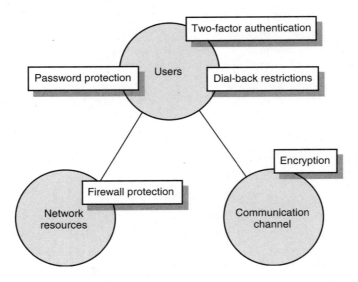

number of questions. Some examples are name, password, home address, mother's maiden name, service request (SLIP, CSLIP, PPP, XRemote, TTY, ARAP, or TN3270), and city of birth. This means that if a user's password is compromised, authentication will not necessarily be granted. The access server's administrator can improve the dialog integrity by routinely changing the questions. The access server's authentication service should be flexible enough to send messages to users' screens. For example, a message might instruct users that their passwords need to be changed because of the company's password aging policy.

The authorization component in access servers allows for greater levels of control over user actions and can be used to create separate administrative groups on the basis of user functionality. For example, a network manager can restrict a user to only perform certain functions on the access server's or router's user interface.

Directory Services

Computer and network administration is incredibly messy. The point nature of IT security means that you must currently separately main-

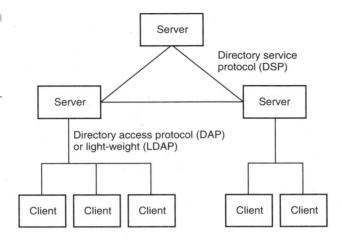

Figure 7-6
Directory Service
Architecture and pro-
tocols (Source: The
Burton Group)

tain router tables, router path management, firewall filters, and access
server privileges—all with separate management tools.

Ideally, a directory service provides a single "place" to register all IT
and network assets in an organization; and makes the process for doing
so formal, controlled, and often automated. Once registered, these assets
can be tracked and managed more easily than the myriad of user access
control lists, VLAN memberships, domains, and rights managers in use
today (see Fig. 7-6).

Once established, the directory becomes a database of all devices and
specific attributes on the network and the relationship between those
devices. It also becomes available for use by other services—including
security services—users, applications, and processes running in the network.

Network Management

Secure network management encompasses a variety of capabilities for
securing the management activity. These include the use of existing
network authentication services and access to the management of network elements and product hooks into standard system administration
security.

Many enterprisewide management platforms are based on UNIX
workstations. These management platforms can be structured to take

advantage of some of the standard system administrator tools available to protect and regulate access in the UNIX environment. System administrators can set up groups with different levels of permission, specifically to control access to any operation that uses the SNMP write privilege community string. For example, a network operations center employee working a late-night shift may not have access at a level allowing reconfiguration, while the manager on the same shift would have this access. In addition, different components such as the router configuration application and the remote LAN access management configuration application can be configured as separate accounts with an additional level of password protection when invoked. This would allow the network configuration to be controlled on a per user basis. The file containing alarm logs, which, for example, would include records of alarms generated by intruders detected by the LAN access management application, can be deleted only by users having "root" privileges (or superuser privileges in UNIX-speak). Using standard UNIX file access facilities, viewing access to this log can also be restricted. In addition, UNIX itself offers a number of ways to add security to a network management console, such as using some strong authentication mechanism for individual user logins (such as SecurID tokens).

A simple but powerful method of providing security within the network management area itself is the network design. When required, it is relatively easy to provide a private network for SNMP traffic only. Using a combination of private segments (i.e., single-user segments), virtual LAN (VLAN) membership, and policy-based routing filters, all agents and management consoles can be effectively isolated from the rest of the network. While vendors promote automatic generation of "communities of interest" based on protocol groupings—through some collective oversight—most have yet to add the SNMP protocol to the list. This would provide the basis of creating an effective, secure, and firewalled VLAN for the network administration/MIS department. This design technique can enhance control for management traffic and agent access to prevent unauthorized users from gaining access to an RMON probe or RMON embedded in a switch (for example) in an attempt to capture and decode live network traffic. Making the SNMP protocol as a community of interest results in all management agents being on a single isolated network, inaccessible—except in special circumstances; for example, the network management team may want to send messages to users on other VLANs—from the management network itself.

IP Security

IP-based VLANs require vigilance by the network administrator. Users can change the IP address configured in their systems, thus altering VLAN memberships. Mechanisms which remember the previous IP address, associated with each switch port, are required to compare changes in IP address and alert the network administrator.

SNMP Security

SNMP, in its present form, is not a secure protocol. For example, SNMP community strings (which are, in effect, passwords) are transmitted as plain text in the SNMP packet. Because of this, many vendors have provided proprietary methods for securing SNMP dialogs between the router management function within the network management router configuration application and the router itself.

Typically, using an initial key (which must be entered at the console of the router initially), dialogs between the configuration application and the router generate encrypted sequence numbers for each request and response in the dialog. The router can conduct multiple encrypted sequence-checked dialogs at once, conversing with the configuration application running at various locations in the network RMON security. Another network element which is a focus of concern for network security is the RMON probe. Some standalone probes minimize security risks of such monitoring devices being used by an unauthorized user. The probe maintains a list of authorized IP addresses. If configured, the probe will not respond to traffic from any other IP address. This includes even a simple PING command. In other words, the probe becomes invisible except to approved management workstations whose addresses are on the list.

Point Products Not Policy

As can be seen, security is not strictly a technology problem—networking vendors these days have vast amounts of security technologies. The real challenge is to enable network managers to implement a single security policy across the enterprise network.

Today's network security market is a hodgepodge of point products: Internet firewalls, dial-in solutions, firewalls, and user authentication for the enterprise and network infrastructure. Implementation of a consistent security policy across the network has been fragmented and labor-intensive.

In February 1997, Cisco Systems announced the Cisco Enterprise Security strategy, the first comprehensive network security initiative proposed by any vendor. This initiative will enable dynamic links between customer policy, user or host identity, and network infrastructure through a cross-product-line integration of available and emerging security technologies. With products scheduled for rollout over the next 18 months, this strategy integrates multiple product lines and takes a fresh approach: an enterprisewide, user-oriented security policy that can be dynamically applied across a network infrastructure. It will transparently integrate current and new standards and technologies, including scalable certification authority, centralized policy management, dynamic access control, and authorization services.

Cisco Enterprise Security is based on three components: identity, integrity, and active audit. *Identity* refers to the dynamic linkage of user authentication, authorization, and location within the infrastructure, allowing use of a single policy for campus, dial, and firewall access. *Integrity* is the feature set that will provide secure firewalls, routing, and device configurations. *Active audit* will enable network managers to detect network anomalies, misuse, and attacks against the network, ensuring that the security policy is both consistent and operating correctly. This new approach to security will enable a consistent, centrally administered policy.

With this security solution, enterprise network managers will be able to allow dynamic setup of user and router access over the Internet through encrypted connections, using IP Security (IPsec), Internet Security Association Key Management Protocol (ISAKMP), and a certificate discovery mechanism based on technology from VeriSign, Inc.:

- Establish an overall enterprise security policy and configure it only once on a centralized server.
- Provide the same authentication and set of privileges to a user regardless of where that user accesses the network.

Cisco is developing products in several phases to support the Enterprise Security strategy. The first of these products will be the *Active Audit Server*, which will provide security audits, verification, and reporting. Second will be the *Enterprise Identity Server*, a multitechnology server for centralized user location, authentication, authorization, and accounting.

This server will be able to dynamically apply a policy to a network infrastructure, using the NETSYS modeling software.

Cisco's Enterprise Security initiative will enable a single policy that can be applied to any potential entry point on the network. In other words, as users access the network from the internal LAN, from the road, or from home, the network knows who they are and what they can do, and network administrators can track it all from one place.

Where We Are Now with VLAN Security

While Cisco's integrated approach to network security is laudable, and will eventually solve the security problems of switched networks and VLANs if fully implemented, we are still at the early stages of VLAN evolution where security mechanisms and security policy (within policy-based VLANs) is not well integrated with the VLAN concept. For example, most VLANs assume static VLAN membership IDs, whereas dynamic VLAN membership IDs would allow a user to move physical location and the VLAN would automatically reconfigure without the network administrator having to do a drag-and-drop operation on the logical network map. There is also the associated problem of associating VLAN membership IDs, MAC address, and network address with various different authentication servers—and possibly different security thresholds for the same person's sets of security and VLAN membership IDs, MAC address(es), and network address(es), one for each authentication server together with a set of passwords to get to an enterprise computing resource—throughout the enterprise, which drives the administration overhead up through the roof.

Currently, virtual LAN security pulls together several aspects of security in a switched environment with the logical domain of VLANs and inter-VLAN traffic. With the use of VLANs, the vulnerabilities of broadcasting are minimized because traffic is partitioned between individual VLANs either by a router or virtual routing function. Routers can be used to isolate broadcast domains contained within sets of switches attached to each router interface. Thus, routing management incorporates access and filtering rules, as has been discussed previously. This "firewalling" is the security baseline for accessing VLANs—a starting point for security policy, not its final expression, as some vendors assume.

Membership of VLANs spanning several switches can be ensured by providing access control for switch ports and associating them with VLAN membership IDs. Such granular access control gives additional flexibility in VLANs which can be centrally administered from a network management console. But again, this should be part of a VLAN security baseline, since most switches provide this facility. It is hardly exceptional.

After being filtered by the switch, the VLAN's user port-MAC-network address can be passed to a central authentication server for further verification of identity, and assigned their VLAN membership ID. Once VLAN users are authenticated to the VLAN, they are assigned authorization parameters which determine where they can go in the network, and what they can do. At the same time, accounting information is recorded for future security audits, utilization analysis, or customer billing. Centralized control simplifies management and helps ensure simple, safe, and consistent administration of policies.

This type of VLAN is known as a *closed VLAN,* because any user who is not on the access control list of the authentication server cannot join the VLAN—no authentication, no membership. This also prevents data "leakage" from a VLAN through promiscuous broadcasts because, by definition, a broadcast from an end station will compromise the security of the closed VLAN—and thus should be automatically blocked by the nearest switch or, failing that, by the inter-VLAN router. Currently, this is the highest available security for VLANs.

The point to note in this discussion is that the policies are consistent for a single VLAN (a simple tautology), but may be incomplete as far as the organization's written security policy is concerned. In some organizations, a single security policy may be applied to all VLANs, because there is no alternative because the vendor's approach to security is not granular enough.

To provide another, higher level of security, inter-VLAN (subnet-to-subnet) traffic can be encrypted at the network level. Encryption at the network level is performed in conjunction with specific protocols (e.g., IP, IPX) rather than specific media, enabling a high degree of flexibility while providing high performance.

Network-level encryption operates on a flow-by-flow basis, encrypting payload traffic between specified user-application pairs, subnets, or VLANs while leaving network layer headers intact. In other words, encryption support is required only at the boundaries of subnets, not at any intermediary networking devices. Because network-level encryption is media- and topology-independent and works well across all interfaces, it is the best approach for large, complex networks (especially those that involve routers or routing functions) and, generally, for networks based on any WAN media.

This simple view of security breaks down when multiple, overlapping, or hierarchical VLANs are considered. If a user is a member of two different VLANs, which VLAN security policy takes precedence? Or does this involve creating a third category of a dual-VLAN security policy containing both policies?

Advanced VLAN Security Topics

Up to this point, this chapter on the security of switched networks and VLANs has focused on automatically tracking computers across a switched fabric with means to identify them as they move and updating policies and access control information.

One of the main claims for VLANs is that they will accurately reflect and support fluid, team-oriented, collaborative processes—the actual virtual workgroup. Virtual LANs must therefore eventually support multiple VLAN memberships since many people may work on several projects at the same time. Consequently, systems which must be part of different VLANs are currently restricted to the single protocol (usually IP, with IPX as the next alternative) they are using. For example, currently a system cannot support multiple IPX or IP addresses—with the exception of multicasting with IP addresses, of course.

DynamicAccess from 3Com, in the form of hardware and software enhancements to its NICs, could add yet another twist to VLAN security. The aim of DynamicAccess is to enable the NIC to interact with the network dynamically, optimize performance for shared and switched networks, participate in RMON data collection, and control desktop multicast traffic. More worryingly, DynamicAccess may also enable user end stations to determine the creation and reconfiguration of VLANs. This would give power to the user in determining which VLANs the user belongs to, and making the appropriate routing decisions. Consequently, it wrecks any centralized administration, since a user can apparently join any insecure VLAN at will.

Supporting multiprotocol virtual LANs is another aspect of multiple memberships. Much of the initial VLAN effort has been IP-centric. In the future, the same benefits of VLANs must be extensible to all common protocol suites. These are inherently different from multiple VLAN memberships since different network layer information can enforce the appropriate separation of broadcast traffic.

Another aspect of VLANs is providing privileges and resources that move with that person (a VLAN user profile) rather than with a particular piece of hardware (the end station). As people move through the enterprise, they should be able to use any computer (the VLAN user profile is device-independent) and carry out activities as if they were using the company desktop in the office or their personal desktop at home (the VLAN user profile is location-independent).

Policy-Based VLAN Security Network Management

VLAN management tools simplify network configuration, VLAN membership management, and overall monitoring tasks. Administrators can reconfigure networks with a simple click of the mouse.

Policy-based management increases network security, enhances network efficiency, is highly scalable, and reduces overall hands-on administration. Further, it helps the move toward managing networks to support business processes rather than simply managing devices.

Policies are established by an administrator at the management station, and administered by a centralized server. This increases network security by centralizing policy establishment. Policies currently available today include

Directory services.   This is not to be confused with network diretory services (see later), since it is based on a limited inventory of users and switches. This registry of users identifies users and their switch ports, providing the route services with information on users not directly connected to the local access switch, reducing the amount of time it takes for a connection to be established.

Security services.   These are based on the information available from the directory services—basically routing restrictions, as opposed to an implementation based on an organization's security policy. This allows the network administrator to limit a user's access to a specified groups of ports and MAC addresses, including address domains. This controls a user's use of network resources, including file and print servers and WAN connections.

Network resource access rules which allow or prohibit workgroup communication or schedule network resources access (e.g., time of day, day of week) can be set up and controlled by an administrator from a central point. This is much easier than programming MAC address filters or protocol-specific access lists.

Note the total lack of integration with point security solutions—this means going back and forth between other security databases and the route-path database of a VLAN and making sure that they match and are up to date. It is another security database to maintain.

Some vendors do not distinguish between switch directory services and security services, but refer to them as *authentication servers*. This does genuine authentication servers a disservice.

Remote Access and VLANs

It is unclear how members of a VLAN in remote and branch offices can be included in a central VLAN. While remote logon to a central server via an authentication server is well understood, its logical equivalent has still to overcome several barriers.

First, because management is central to virtual LANs, VLAN management capabilities must be extended outward to encompass the remote office. As a first stage, remote stackable hubs should be replaced with stackable switches with integral VLAN support for the remote office. This pushes VLANs out into the wide area where link speeds are a fraction of a fast LAN backbone or even PC-switch link and cost is at a premium. This will restrict the attractiveness of strategic VLANs.

A further problem is that, to date, all VLANs assume a continuous connection between the PC and the switch. To overcome this, switch vendors will need to develop the concepts of dynamic VLAN membership further to include intermittent connections and multiple VLAN memberships—for example, a salesperson in the sales VLAN will have one (static) address while sitting at the office and a different (dynamic) address when logging in remotely from the laptop. The implication for authentication servers is that they must accept both addresses as legitimate, but not both at the same time.

With remote users, the assignment of the VLAN IP address must be made after the user has been authenticated as a VLAN user on the central VLAN. This means that authentication access servers of all kinds must be extended to include tables of valid VLAN members. In the case of highly mobile users, PAP/CHAP authentication needs to be mapped to VLAN memberships.

Single Signon, Directory Services, and VLAN Membership

In the directory services database (as mentioned in Chap. 4), there is an obvious relationship between items in the database and the right of access by any given item has to another. Users, processes, and applications may need one or two levels of access to network resources—either administrative access or a more restricted level suitable for the accomplishment of day-to-day tasks (see Fig. 7-7).

The aim is to incorporate all network resources and access points into a single administrative and security model and to enable controlled

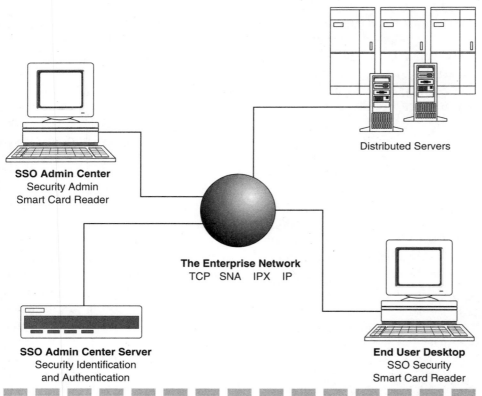

Distributed Servers

SSO Admin Center
Security Admin
Smart Card Reader

The Enterprise Network
TCP SNA IPX IP

SSO Admin Center Server
Security Identification
and Authentication

End User Desktop
SSO Security
Smart Card Reader

Figure 7-7   Single signon: systems, networks, and services in the enterprise (Courtesy of Unisys Corporation, 1977)

access through a single login process. Instead of logging into, say, the finance VLAN, users should need to log in only once to directory, and then directory service will handle the authentication they require to access all other services, VLANs, and applications transparently. To do this, users will need two forms of directory information to be distributed and synchronized across switches and servers:

- *Identity*—switches and servers will check user and application identity by examining digital certificates.
- *Policy*—all users and applications will have policy tightly linked to their identities—explaining where they can go in the network, how much bandwidth they can reserve, and what information they can see.

Directory services can therefore hand out policies—including security policies—to the network, not network management, as with current policy-based VLANs. Directory-driven switches will enable MIS to administer the network, workstations, and servers with a single set of policy tools. With digital certificates and policies controlling the universe, users will get a single logon, and MIS can get rid of multiple password and ID administration, including VLAN membership ID administration.

For directory services to take over the world, intelligent switches, routers, and hubs must be directory-enabled and "directory-aware." Cisco is expected (at time of writing) to deliver its directory services during 1997—and add directories and directory management services, called *DistributedDirector*, to its switches and routers.

Directory-enabled switches will understand where users and applications are able to go. MIS can stop copying databases in and out of separate authentication servers and messing around with passwords and aligning VLAN membership IDs with remote-access IDs and passwords, over and over.

Directory servers (whether Novell NDS, Netscape, Microsoft Active Server, or Cisco's DistributedDirector) can deliver policies to intelligent switches. With embedded policies, switches will know what quality-of-service level should be applied to a one-off IP video session and what priority it gets from the network in terms of bandwidth made available to it. Directory servers can propagate directories and (security) policies throughout the network.

Security Standards

ISAKMP and Oakley Protocols

The ISAKMP and Oakley protocols are leading contenders for Internet key management being considered by the IP Security (IPsec) Working Group of the IETF.

ISAKMP. The Internet Security Association and Key Management Protocol (ISAKMP) provides a framework for Internet key management and provides the specific protocol support for negotiation of security attributes. By itself, it does not establish session keys; however, it can be used with various session key establishment protocols, such as Oakley, to provide a complete solution to Internet key management.

Oakley. The Oakley Key Determination Protocol uses a hybrid Diffie-Hellman technique to establish session keys on Internet hosts and routers. Oakley provides the important security property of Perfect Forward Secrecy and is based on cryptographic techniques that have survived substantial public scrutiny. Oakley can be used by itself, if no attribute negotiation is needed, or in conjunction with ISAKMP. When ISAKMP is used with Oakley, key escrow is not feasible.

ISAKMP-Oakley Hybrid. The ISAKMP and Oakley protocols have been combined into a hybrid protocol. The Resolution of ISAKMP with Oakley uses the framework of ISAKMP to support a subset of Oakley key exchange modes. This new key exchange protocol provides optional Perfect Forward Secrecy and full security association attribute negotiation, as well as authentication methods that provide both repudiation and nonrepudiation. Implementations of this protocol can be used, for example, to establish virtual private networks (VPNs) and also allow remote users from remote sites (who may have a dynamically allocated IP address) to access a secure network.

ISAKMP software—and hence the hybrid protocol—is export-controlled and is not generally available outside the United States and Canada unless certain specific criteria have been met. Commercial copies are available from MIT (within the United States and Canada).

US DoD Implementation

The U.S. Department of Defense, Office of Information Security Research, has made its ISAKMP Prototype Implementation freely available for distribution within the United States. This implementation does not include any session key exchange capabilities, but does include full ISAKMP features.

RADIUS

Livingston Enterprises developed a distributed security solution called *Remote Authentication Dial-in User Service* (RADIUS), which solves the problems associated with meeting the security requirements of remote computing. This solution eliminates the need for special hardware and provides access to a variety of state-of-the-art security solutions. Distributed security separates user authentication and authorization from the communications process and creates a single, central location for user authentication data.

Based on a model of distributed security previously defined by the Internet Engineering Task Force (IETF), RADIUS provides an open and scalable client/server security system. The RADIUS server can be easily adapted to work with third-party security products or proprietary security systems. Any communications server or network hardware that supports the RADIUS client protocols can communicate with a RADIUS server. Livingston offers the RADIUS server free of charge to its customers and supports the RADIUS client protocols in its PortMaster family of communication servers and routers. Client/Server Architecture RADIUS is a system of distributed security that secures remote access to networks and network services against unauthorized access. RADIUS includes two pieces: an authentication server and client protocols. The server is installed on a central computer at the customer's site. RADIUS is designed to simplify the security process by separating security technology from communications technology. All user authentication and network service access information is located on the authentication, or RADIUS, server. This information is contained in a variety of formats suitable to the customer's requirements. RADIUS in its generic form will authenticate users against a UNIX password file and Network Information Service (NIS), as well as a separately maintained RADIUS database.

Communications servers working with modems—such as the PortMaster—operate as RADIUS clients. The RADIUS client sends authentication requests to the RADIUS server and acts on responses sent back by the server.

How RADIUS Works.   RADIUS authenticates users through a series of communications between the client and the server. Once a user is authenticated, the client provides that user with access to the appropriate network services. The following is a description of the authentication process using a PortMaster Communications Server and RADIUS.

Using a modem, the user dials in to a modem connected to a PortMaster Communications Server. Once the modem connection is completed, the PortMaster prompts the user for a name and password. The PortMaster creates a data packet from this information called the *authentication request*. This packet includes information identifying the specific PortMaster sending the authentication request, the port that is being used for the modem connection, and the user name and password. For protection from eavesdropping hackers, the PortMaster, acting as a RADIUS client, encrypts the password before it is sent on its journey to the RADIUS server.

The authentication request is sent over the network from the RADIUS client to the RADIUS server. This communication can be done over a local or wide area network, allowing network managers to locate RADIUS clients remotely from the RADIUS server. If the RADIUS server cannot be reached, the RADIUS client can route the request to an alternate server.

When an authentication request is received, the authentication server validates the request and then decrypts the data packet to access the user name and password information. This information is passed on to the appropriate security system being supported. This could be UNIX password files, Kerberos, a commercially available security system, or even a custom-developed security system.

If the user name and password are correct, the server sends an *authentication acknowledgment* that includes information on the user's network system and service requirements. For example, the RADIUS server will tell the PortMaster that a user needs TCP/IP and/or NetWare using PPP (Point-to-Point Protocol) or SLIP (Serial Line Internet Protocol) to connect to the network. The acknowledgment can even contain filtering information to limit a user's access to specific resources on the network. If at any point in this log-in process conditions are not met, the RADIUS server sends an *authentication reject* to the PortMaster, and the user is denied access to the network. To ensure that requests are not responded to by unauthorized hackers on the network, the RADIUS server sends an *authentication key,* or signature, identifying itself to the RADIUS client. Once this information is received by the PortMaster, it enables the necessary configuration to deliver the right network services to the user.

8

Benefits, Problems, and Performance Considerations

Benefits of VLANs

A VLAN objective often mentioned by forward-thinking vendors is the establishment of the virtual workgroup, where members of the same department of an enterprise can all appear to share the same LAN, with most of the network traffic therefore staying within the same VLAN broadcast domain. Someone moving to a new physical location but remaining in the same department could move without involving workstation reconfiguration. Similarly, users would not need to move physically if they changed departments—the network manager would change the VLAN membership instead. This idea supports the concept outlined in Chap. 2 of fluid project-based teams of workers virtually connected to the same LAN and dynamically assigned to VLANs for the duration of the project while allowing users to remain in the same physical locations.

Unfortunately, attractive as this scenario is, VLANs alone are not all that is needed to implement it—there are managerial issues of control and culture. After all, people choose to be physically located near workmates rather than to add to network efficiency. Their managers also like to be able to physically keep an eye on them, although managers are having to cope with less control as staff work in more flexible ways, anyway. If such fluid and transitory teams become the norm, then the constant readjustments will make nonsense of the idea of reducing the cost of moves, adds, and changes in networks—a primary benefit of VLANs often cited. Updating VLAN membership could become as onerous as updating routing tables is today, although it may save on the time and effort involved in physically moving the user's workstation.

To reiterate a point already discussed briefly in Chap. 4, VLANs as they stand today should really be seen as a solution to at least one of two problems:

- Containment of broadcast traffic to minimize dependence on routers

- Reduction in the cost of network moves and changes

Eventually, with the more advanced policy-based VLANs, it will be possible to do this and much more: to set filters and membership IDs in such a way as to make the network reflect the business and be more secure.

The IEEE 802.1Q standard sets out the following possible benefits of VLANs:

- VLANs facilitate easy administration of logical groups of stations that can communicate as if they were on the same LAN. They also facilitate easier administration of moves, adds, and changes in members of these groups.

- VLANs are supported over all 802 media, and over shared media as well as point-to-point LANs.

- Each distinct VLAN is uniquely identified throughout the bridged LAN. There is a consistent representation of a VLAN across a VLAN fabric (also across ATM), so that the shared VLAN knowledge of a particular packet remains the same as the packet travels from one point to another in the VLAN fabric.

- In the absence of VLAN configuration, bridges will work in plug-and-play 802.1D mode. End stations will be able to communicate throughout the bridged LAN.

Each of these points is further examined below, but it should be noted that all of these aimed-for benefits still consider only the efficiency of the network and not any of the more sophisticated policy-based benefits that will come along in time.

Containment of Broadcast Traffic

Although switches and bridges enhance performance compared with routers, layer 2 switches do not filter LAN broadcast traffic, instead replicating it on all ports. This not only can cause large switched LAN environments to become flooded with broadcasts but also wastes WAN bandwidth. As a result, users have traditionally been forced to partition their networks with routers that act as broadcast "firewalls."

The primary benefit of VLANs, given a switched network, is that LAN switches supporting VLANs can be used to control broadcast traffic, reducing the need for routing to the edge of the network and to move inter-VLAN traffic. Broadcast traffic from servers and end stations in any particular VLAN is replicated only on those switch ports connected to end stations belonging to that VLAN. Broadcast traffic is blocked from ports with no end stations belonging to that VLAN, in effect creating the same type of broadcast firewall that a router provides. Only packets that are destined for addresses outside the VLAN need to proceed to a router for forwarding.

In a switched IP environment without VLANs, broadcast traffic will cause enormous congestion. Networks with switches need dividing logically into VLANs to give the benefit of broadcast containment as the VLANs can limit broadcasts to prevent them from propagating across the entire network. Traffic between VLANs is then firewalled by routing, and this limits the propagation of multicast and broadcast traffic between VLANs.

Related to the containment of broadcasts is the reduction in error propagation. End-station errors stop at the switch rather than propagating throughout the network with concomitant snowball effect. This allows a network manager to bring up thousands of users at once without worry.

Cost Reduction

Reducing the Cost of Moves, Adds, and Changes.   Administrative management and operations are usually the main expenses over the life cycle of networks, while capital costs such as the purchase of hardware and software represent an increasingly smaller proportion, usually under 20 percent. As business processes flatten and move toward cross-disciplinary, fluid project teams (as set out in Chap. 2), the traditional ways of handling moves, adds, and changes manually are not effective or efficient. Annual change rates of over half the numbers of end stations is not uncommon, and if the process were easier, there would likely be more changes even than this.

The cost of network management can be brought down in several ways. Here the question of the network managers' time is considered, as it is possible, if the VLAN is chosen for this reason, for network managers to spend less time configuring or reconfiguring the network. Instead of needing perhaps two staff members dedicated to supporting every 1000 ports, one can manage up to 5000 ports.

The main way in which time can be saved is by reducing the cost of moves, adds, and changes, at present about 66 percent of the IT budget. As pointed out in Chap. 4, installing a VLAN does not necessarily save time—it depends on the type chosen. All VLANs decouple the configuration of the network switches from the endpoint station network layer or MAC address. Workstations and routers no longer require reconfiguration when users physically move from one location to another. Types vary; as described in Chap. 4, some are time-consuming to configure initially, but then changes are automated; others are set up automatically, but later changes must be done manually. VLANs assigned to switch ports or MAC addresses are easy to set up and are preferred by many users whose network consists of up to a few thousand end stations, although these are not scalable and changes must be made completely manually in the case of port switching. The advantage of layer 2 MAC-address-based VLANs is the rapid switching speed of ASIC-based switches rather than these change configurations.

Normally, when a user moves to a different subnet, IP addresses must be manually updated in the management workstation. This updating process can consume a lot of time that could be used for more productive endeavors. VLANs eliminate that updating process because VLAN membership is not tied to a workstation's location in the network, allowing moved workstations to retain their original IP addresses and subnet membership. Fully automated-throughout IP-based VLANs are possible, but troubleshooting these is usually awkward as the IP address allocation process has first to be unraveled. If saving administration time is the main reason for installing a VLAN (after the necessary broadcast containment), then the type should be chosen carefully, but it is possible to save a lot of time this way.

Other Cost Advantages. When compared to traditional routers, virtual networking provides greater performance at lower cost. The purchase price of a switch can be one-tenth the price of a traditional router based on the same packet throughput. Switches with routing functions on them are more expensive than simple layer 2 switches, but are still cheaper than the large routers usually used to segment LANs before switches were employed.

Operational costs will also be less because of lower administrative overhead quite apart from costs of moves, adds, and changes—routers are notoriously easy to misconfigure and need more maintenance. Reliability of the network is improved through the connection-oriented transport service, automatic fault detection, and connection rerouting capabilities.

Network Advantages

The best switches available on today's market have four components to their success:

- Smart silicon delivering high bandwidth
- Steadily decreasing prices
- Intelligence to support management functions
- Management software allowing the efficient and cost-effective administration of these new environments.

Switching has to coexist with the installed base of shared-media LANs and wide area backbone technologies in many networks, at least during migration. It is here that the 802.1Q's insistence that bridges will work in plug-and-play 802.1D mode in the absence of VLAN configuration comes into its own. End stations will be able to communicate throughout the bridged LAN, with default VLAN assignment as data enters the VLAN zone.

In a traditional network broken into broadcast domains by routers, expansion means an increase in the number of routers until latency begins to degrade network performance. This is bad enough for older applications, but is especially bad with the newer ones, which feature delay-sensitive multimedia and interactivity. Switches that employ VLANs can accomplish the same division of the network into broadcast domains, but can do so at latencies much lower than those of routers (this varies with each vendor). Switch performance, measured in packets per second (pps), is higher for switches than for traditional routers. At present, the fastest routers (from Bay Networks and Ascend) runs at about 1 million pps; the fastest IP switch (from Digital Equipment) is claimed to run at 20 million pps—although just how this is measured has not been revealed. However, it should be noted that some layer 3 switches may not be faster than most routers. However, performance, whether in terms of sheer speed or of latency, is related to the number of hops traversed—for any very large LAN or a VLAN extended across the WAN, performance varies with the number of hops.

Management Tool Integration

If information from the VLAN management system can be integrated with the enterprise management environment, administrators can have a complete view of the behavior of switching fabrics in relation to shared and switched LANs and WAN technologies. If integration with other management tools is also provided, many advantages can be conferred; for example, trouble-ticketing systems can be integrated so that VLAN management tools automatically create and populate a trouble ticket and allow staff to track a problem through to resolution.

Keeping the 80/20 Rule

Virtual LAN support for virtual workgroups is often tied to support of the aforementioned 80/20 rule, that is, 80 percent of the traffic is local to the workgroup while 20 percent is remote from or outside the workgroup. By configuring VLANs to match workgroups, the theory is that only the 20 percent of the nonlocal traffic will need to pass through a router and out of the workgroup, improving performance for the other 80 percent of the traffic. This is a benefit in those networks where the 80/20 rule still holds, but as has been said already, nowadays far more of the traffic goes out of the workgroup than in days of yore, with the introduction of intranets, email, groupware, and client/server environments.

Efficient IP Addressing

VLANs also have another advantage; they make efficient use of scarce IP addresses. Routers use layer 3 addresses to move data between LANs. In general, layer 3 addresses (subnet/network numbers) identify the broadcast domains or bridge domains where the frame's destination device can be found. In the case of IP, a network administrator assigns an address and enters it manually into each workstation's configuration, or the address assigned dynamically through a DHCP server. Each network address identifies a workstation and the LAN segment where that workstation can be found.

Theoretically, network addresses can be assigned any way that an administrator chooses. Unfortunately, that's not true in practice for most networks today. The reason is the Internet. To mesh a private network with the Internet, most organizations use network numbers assigned by the Internet Number Authority. The explosive growth of the Internet has quickly used up a large percentage of the possible network numbers, and the numbers available to organizations has become very limited (unless they decide to use the IP addresses out of the private address space in combination with network address translation). Routers do not help administrators conserve subnet numbers since each port on a router is generally its own subnet.

Virtual LANs reduce the waste of scarce IP addresses by using a limited number of network addresses very efficiently. In a hub/router-based network, each LAN segment usually needs its own subnet number so that routers can move data between subnets. This quickly becomes inefficient unless there are enough users on a single router port to fill in all the possible IP addresses. Using variable-length subnet masks helps preserve the IP address space to a certain extent in hub/router-based networks. In a network based on virtual LANs, any number of LAN segments can be combined into a single virtual LAN, allowing administrators to assign IP addresses and subnets more efficiently. A subnet can be one port on one switch, multiple ports on one switch, or multiple ports on multiple switches.

Benefits of Overlapping VLANs

For such purposes of sharing printers, and also to enable network managers, for example, to oversee all VLANs, it is often a good idea to place such resources in overlapping VLANs, although at present this limits the VLAN type to a MAC-address one, and this is therefore not scalable. The overlapping VLAN method enables, for instance, the printer in the sales and marketing area to be used by other groups, such as accounts, which may not be on the same VLAN although physically located nearby. If all printers are on all VLANs, with default printers set up for each VLAN but others available by choice, then in the case of printer malfunction, all users can print, even if it means a walk. This obviates the need for the print file to traverse a router connecting the VLANs. Also, it is useless to use a VLAN to access a single server [whether mainframe, VAX (Digital's Virtual Access Extension), or UNIX cluster] if that server is accessed by multiple points throughout the enterprise. Likewise, it is important to integrate the process—not just the technology—to reap the benefits of the virtual approach.

Policy-Based VLAN Benefits

Another great benefit, mainly yet to be realized, will be the incorporation of policy-based management. As mentioned in Chap. 4, VLANs thus created are managed with software which allows for the incorporation of rules and constraints to control and direct the management system. Management tasks are then carried out to satisfy policy rules supporting the business processes of each client. Policies can identify critical applications and ensure that users continue to receive the bandwidth they need when failures occur. Using a combination of private segments (i.e., single-user segments), virtual LAN (VLAN) membership, and policy-based routing filters, all agents and management consoles can be effectively isolated from the rest of the network.

As policy-based VLANs are implemented, routing policies will be brought in early on, allowing administrators to assign bandwidth and priorities to specific applications and to connections between VLANs or connections within VLANs. If the VLAN is policy-based, then each user's policy of use can be attached to the network ID instead of the hub or router port. Each user's packet flow can then be traced through the network, due to the inherent connection-oriented service, allowing protocol-related problems to be resolved quickly. There is no need to predefine VLANs for endpoint stations using the supported layer 3 protocols.

Policy-based VLANs can offer better network security. Consider a network built with hubs. Anyone can plug a protocol analyzer (or a PC with protocol analyzer software) into any connection on a hub and intercept all the data that is being sent on that segment. The data being intercepted could easily be confidential. If each device is connected to its own port on a network with switched VLANs, then that's not possible. The only observable information is intended for the device attached to that port. With policy-based VLANs, MAC and IP addresses can be bound to a port—ensuring that only users with a specific MAC and IP address can operate on a switch port adding to a user's security.

All this sounds ideal, and could be so, but it must be recognized that setting up a policy-based VLAN needs careful thought at the inception if benefits are to be realized. The processes are complex:

- Policies must be set per department and/or per user.
- Policy layers must be organized.
- Applications may need to be brought within that policy base.
- Mobile users may require a temporary change in policy.

On the last point, for example, visitors from other parts of the enterprise should be able to sit down at any computer, properly identify themselves, and have their normal virtual LAN access to resources, while other members of the VLAN are constrained by normal policies.

Problems with VLANs

When VLANs are deployed in existing networks, they can affect the performance and reliability of other applications in ways which are difficult to predict, understand, isolate, and solve. Problems can range from graceful degradation, to intermittent failure, to disaster. If the network infrastructure moves from router-based to switch-based, this compounds problems as network support staff gradually familiarize themselves with new technology.

Interoperability

As pointed out many times so far, VLANs have their downsides as well as upsides. The most significant weakness is that VLANs have been, to date, single-vendor solutions and therefore may lead to switch vendor lockin. Apart from the fact that one has to buy whole VLAN schemes from the chosen vendor; one cannot pick and choose bits here and there to get the desired functionality; the schemes differ in the way they work at the lowest level. For example, vendors differ in the way they solve the problem of occasionally exceeding the maximum length of MAC-layer frames as these headers are inserted (a problem with 802.1Q as mentioned earlier). It should by now be apparent that no one vendor has the total solution, but no two vendors can guarantee interoperability. There can be interoperability problems due to

- Having equipment from different vendors
- Having hardware from a vendor that does not work with the VLAN software
- Having management software from a vendor different from the VLAN software vendor

If neither broadcast containment nor reducing the cost of moves, adds, and changes is a problem, then the user organization may want to put off installing VLANs until they are more standard and interoperable and continue deploying a multivendor network backbone, segmented by a mix of a few routers and a relatively large number of simple switches.

Discarding Equipment

The cost implications of junking perfectly good equipment because it could not handle the VLAN are considerable. All VLAN specifications involve tagging packets with an extra header. Many ASIC-based switches will have to be thrown out when implementing VLANs under any standard or protostandard according to Mick Seaman, chair of the 802.1Q interworking task force. Even though CPU-based switches could be upgraded in software, handling the extra header could burden the switch CPU, causing a dropoff in performance. Some vendors don't want to take the hit. For example, Cisco could upgrade its software but has decided instead to roll out new hardware. "Software would be easier, but performance means it makes more sense to do this in hardware," says Jayshree Ullal, director of marketing with Cisco's Workgroup business unit. If you are buying switches, it is best to check that they are VLAN-enabled; if buying a VLAN, check that any of your existing switches can handle both the extra tag and the performance needs.

To maintain VLAN boundaries strictly, VLAN boundaries legacy internetworking equipment should be very carefully placed; if not, you could end up bridging or routing between VLANs with legacy equipment. Depending on the complexity and demands of the network, this aspect should be looked at closely. This could have an impact on performance, security, and management time. It may be best to replace existing routers as well as switches, despite the migration strategies recommended by some vendors, for example, Bay Networks, where gradually bringing over users into VLANs is a supposed advantage. Existing bridges act exactly as switches under the first iteration of 802.1Q, which deals only with bridges (including switches as much the same animal). This necessitates the ability to handle larger frames than normal and in future, multiple spanning trees (see below in section on standards problems); this will be a problem for legacy bridges.

A Necessity or a Band-Aid?

It has even been said that VLANs are just a solution looking for an application—this is not true wherever there are switched networks, unless broadcast storms are welcomed, but it is possible for a skeptic to ask whether VLANs exist because they really are useful, or whether they help cover up other faults in the network. This is the key question, especially in the following cases:

- For broadcast issues, shouldn't one deal with the source of broadcasts in a more fundamental manner, if they are a problem, such as finding ways to limit the production of broadcast traffic?

- For security issues, shouldn't one use encryption between communicating parties rather than introduce a potentially large number of possibly vulnerable access points via the more complex switch/VLAN environment?

- For network management complexity, shouldn't proper business practices be worked out rather than giving managers tools that may cause them more, not less, work?

The limitations of the different types of VLANs vary, and each is discussed below. It should be noted, however, that the only drawbacks specific to policy-based VLANs, apart from their proprietary nature, are their scarcity.

The disadvantages of any layer 2 VLAN (port-based or MAC-address-based) is that it is limited to flat networks and it is not highly scalable to accommodate networks with tens of thousands of users. Another disadvantage of VLANs is that communication between VLANs requires routing using a network protocol. Routing adds expense and cripples performance. Also, VLANs do nothing to address the operational costs associated with conventional routers. A severe limitation of layer 2 VLANs is that they do not support end-to-end communication between LANs connected by a WAN.

Port-Based VLANs

The main problem with port-based VLANs is that the network manager must reconfigure VLAN membership when a user moves from one port to another, thus negating the benefits of easing moves, adds, and changes to a network. Attaching a system to a different switch port requires an interaction with the management system to move the system's membership to a new port. Similarly, another interaction is necessary if the membership is changed without moving the system. So, while port definition is simple and straightforward, it limits user mobility since users cannot move if it means a lot of work for someone else.

Security concerns also arise since a system on a closed, private VLAN can be replaced and an intruder substituted in its place. The intruder receives broadcast traffic flowing within that virtual LAN.

Both port-based and MAC-address VLANs do the job of confining traffic to separate virtual domains, but, as seen in Chap. 4, port-based VLANs require manual operation throughout and MAC-address VLANs need all original configuration details entered by hand. And both need a router to pass VLAN traffic between the separate VLANs, with a big performance hit associated. Table 8-1 summarizes the benefits and limitations of the various VLAN types.

MAC-Address-Based VLANs

Layer 2 VLANs provide some value over traditional LANs by limiting broadcast and multicast traffic to only those users of the same administratively defined VLAN group. However, this value comes at the relatively high cost of administrative support, despite being designed to make this easier. Users must be moved from VLAN to VLAN or be placed on more than one VLAN. But placing users on more than one layer 2 VLAN defeats the whole concept of VLANs in the matter of broadcast containment, despite the usefulness of overlapping VLANs as noted above, and it means even more time spent at the initialization of the VLAN system. It is also not possible to have this feature on a layer 3 LAN, as the MAC addresses are not used across the network layer.

MAC-address-based VLANs are best within a totally switched environment, especially with a single user per switched port, but this cannot always be the case in the real world. Where they are implemented in

TABLE 8-1

Sizing up the VLAN Types

Features	Port-based	MAC-based	Layer 3—based	Protocol-based	Multicast	Policy-based
Flexible	No	Moderate	Moderate	Moderate	Moderate	Very
Easy to assign	Yes	No	Variable	Yes	Variable	Yes
Device movements tracked	No	Yes	Yes	Yes	Yes	Automatic
Use of network structures	No	No	Yes	No	Yes	Automatic
Multiple virtual LANs on a single port (hub/server) support	No	Yes	Yes	Yes	Yes	Automatic
One device in multiple virtual LANs	No	Possible	Possible	Possible	Possible	Automatic
Security	High	Minimal	Minimal	Minimal	Minimal	Selectable

Source: Xylan Corporation.

shared-media environments, they will run into serious performance degradation if members of different VLANs coexist on a single switch port. In addition, the primary method of communicating VLAN membership information between switches in a MAC address-defined VLAN also runs into performance degradation with larger-scale implementations. But then, as pointed out in Chap. 4, the MAC address VLAN schemes are not scalable for many reasons.

A major limitation is the cost to network management time—they are usually only semiautomated, and this limits scalability. Also, the inability of layer 2 VLANs to span traditional routing domains has limited the deployment of VLANs. Without support for layer 3 routing and policy-based management, the value of the VLAN concept to an organization is unclear. The inability to span traditional routers has limited layer 2 VLAN deployment as a workgroup solution. Conversely, semiautomatic VLANs share this limitation; while tools are there for the initial setup, this type can have problems with later changes.

As pointed out earlier, one drawback of MAC-address-based VLAN solutions is the requirement that all users initially be configured to be in at least one VLAN. The disadvantage of having to initially configure VLANs becomes clear in very large networks where thousands of users must each be explicitly assigned to a particular VLAN. Some vendors have mitigated the onerous task of initially configuring MAC-based VLANs by using tools that create VLANs based on the current state of the network—that is, a MAC-address-based VLAN is created for each subnet.

Another, but minor, drawback to VLANs based only on MAC-layer addresses emerges in environments that use significant numbers of notebook PCs with some docking stations—very common in the types of workplaces described in Chap. 2. The problem is that the docking station and integrated network adapter (with its hard-wired MAC-layer address) remains on the desktop, while the notebook travels with the user. When the user moves to a new desk and docking station, the MAC-layer address changes, making VLAN membership impossible to track.

IP-Based VLANs

An IP-based VLAN definition requires consideration; although these make the best automated VLANs, when things do go wrong, there is an extra layer of complexity to unravel before it can be put right, meaning more work for the hard-pressed network manager—and it is easy for users to change the IP address configured in their systems, thus altering their VLAN memberships. Mechanisms to remember the "state" associated with each port are required to note a change in IP address and alert the network administrator if this is not to be a security risk as well as a disturbance to the system.

If a device moves from one location to another and it was manually configured, then a network administrator needs to change that workstation's network address as well as its gateway information. Dynamically assigning IP addresses through a DHCP server resolves the workstation configuration issue but makes it almost impossible to implement deterministic security policies in enterprise networks. Since DHCP assigns an IP address on the basis of the subnetwork number that the router port belongs to, whenever a workstation moves to a new subnetwork, layer 3 policies often become invalid and new firewall rules have to be defined, adding more complexity to network management. This isn't much work if this seldom happens, but in a large network, moves and changes are quite common. The network change process can eat up quite a bit of an administrator's time and cost a company a significant amount of labor time (and money). Until an administrator changes a relocated workstation's address or modifies the layer 3 security policies (sometimes at many locations in the network), that workstation cannot communicate with the network.

Increasing Network Management Complexity

VLANs are supposed to make a network manager's life easier but, as has been noted, they may actually increase the complexity. Instead of making moves, adds, and changes easier, they can be made harder, given the increasing numbers of these the network manager may be expected to cope with unless the VLAN is of an automatic type. Apart from this increase in complexity, management of the network itself, apart from the VLANs, may be made more difficult. For example, if a LAN manager restructures a Novell NDS tree or a set of Windows NT server domain/trust relationships due to moves, adds, or changes, VLANs now double the management workload. The network manager now must keep the VLAN structure consistent with the network OS directory structure.

As pointed out at length in Chap. 6, implementing virtual LANs means changing the rules for network and system administrators. Unfortunately, the necessary changes may not be integrated with the network and systems designers, nor with other software. Synchronization of the layout of servers so that they mesh and maximum availability is important. The network management side is not pretty. Unfortunately, the ubiquitous SNMP was developed to manage physical network devices— not VLANs. It is not a simple task to integrate VLANs with SNMP, as seen in Chap. 6. The RMON protocol looks at LAN segments and can monitor the health of groups of network devices; what RMON needs for VLANs is the ability to dynamically manage each virtual port of a virtual physical layer switch, which involves costly CPU overheads.

ATM management, as we have seen, is even more of a nightmare. It doesn't yet integrate with SNMP (the ATM Forum has only just begun to deal with congestion control and MIB integration with SNMP), and it is difficult to manage the integration of ELANs and VLANs. It is not unusual to find different traffic management tools for ATM and Ethernet switches, each producing totally different sets of statistics which are difficult, if not impossible, to correlate in any meaningful fashion. There is also the problem of scalability, again. SNMP managers have problems managing hundreds of events—let alone correlating those network events. In the world of ATM, networks will support several thousand users. For ATM, too, there are no testing devices for speeds of

≥155 Mbits/s and, as said earlier, you cannot manage what you cannot measure.

Standards Problems

The main problem with the standard is that it is not here yet but VLANs are already being employed as a necessity in the switched networks now being set up.

The first version of the 802.1Q standard will not be the definitive one, only a first stab at the solution. There are several omissions. For a start, the additional header in the 802.1Q standard can cause illegally long Ethernet frames (giants) compared with 802.3. If frames are too long, they're going to show up as errors on an SNMP management console. They may be dropped, depending on the switches installed, or they may cause no problems, at least most of the time, but the management hit is there. If they are dropped, then this will cause a lot of performance problems and may produce real loss of data eventually. The solutions being considered by the standards committee include

- Increasing Ethernet frame size
- Adding software to end stations to restrict length

But it is clear that in the first iteration the problem will not be fixed, and the next time around it may mean another upgrade (forklift or otherwise) to the switches.

A further problem: The 802.1Q standard supports only a single spanning tree even across multiple VLANs in this first version, although for performance needs and scalability, it is obviously better to take full advantage of a switch's capacity to use many paths; once again this leads to a performance hit.

Leaving aside 802.1Q and looking at the "other" VLAN standard—802.10, which is really for tagging for security purposes, it is possible that a switch will attempt to interpret security information as a VLAN address, with results of nondelivery and SNMP faults, or incorrect delivery (by coincidence) and security leaks. The variable-length field also causes performance problems. There is also, of course, the problem of being unable to conjoin VLANs using the two different standards 802.10 and 802.1Q, a situation that can arise if one company buys another or if two previously separate parts of a company without a central buying policy become part of an intranet.

A further problem can arise with the well-known and widely implemented spanning-tree algorithm—the complexity of management of VLANs can mean that spanning-tree changes can increase the long time (in some cases to well over a minute) needed to recalculate paths if a switch drops out of the network. This is exacerbated by the length of time the software can take to respond to management requests; again, a problem caused by the complexity of the management.

Security Problems

As pointed out earlier, we are still at the early stages of VLAN evolution where security mechanisms are not well integrated with the VLAN concept. For example, most VLANs assume static VLAN membership IDs, whereas dynamic VLAN membership IDs would allow users to change their physical locations and the VLAN would be able to automatically reconfigure without requiring the network administrator to perform a drag-and-drop operation on the logical network map. There is also the associated problem of associating VLAN membership IDs, MAC address, and network address with various different authentication servers—and possibly different security thresholds for the same person's sets of security and VLAN membership IDs, MAC address(es), and network address(es), one for each authentication server together with a set of passwords to get to an enterprise computing resource—throughout the enterprise, which drives the administration overhead up through the roof.

Unexpected events can compromise VLAN security. For instance, if the ports on a switch have been set up to customize VLAN membership, and a power outage occurs, the switch may come back up with its default setting (if, say, it can't find the configuration file holding the programmed information). It could therefore run with these defaults and not segment the traffic according to the defined VLANs. Traffic which previously had been split may not go in the required directions until the switch is again reconfigured as wanted. This may have no more consequence than inconvenience and poor performance (if all the defaults are that everything is in one big VLAN), or it could have serious security consequences with sensitive information going to the wrong addresses.

IP-based VLANs require vigilance by the network administrator. Users can change the IP address configured in their systems, thus altering VLAN memberships. For example, 3Com's DynamicAccess may enable user end stations to determine the creation and reconfiguration of

VLANs. This gives a user power to determine which VLANs he or she belongs to and to make the appropriate routing decisions. Consequently, it could wreck any centralized security administration, since a user can join a VLAN at will. Mechanisms which remember the previous IP address, associated with each switch port, are required to compare changes in IP address and alert the network administrator.

SNMP, in its present form, is not a secure protocol. For example, SNMP community strings (which are, in effect, passwords) are transmitted as plain text in the SNMP packet. Because of this, many vendors have provided proprietary methods for securing SNMP dialogs between the router management function within the network management router configuration application and the router itself. While proprietary add-ons are often a good way of obtaining extra functionality, if they conflict with a new standard or proprietary software tool, such as VLANs, then security may be breached, and a problem such as this may not always become apparent. A malevolent or mischievous user who gains access to password strings can wreak havoc. As we have seen, most vendors have yet to add the SNMP protocol to the list. If they did, this would provide the basis of creating an effective, secure, and firewalled VLAN for the network administration/MIS department.

So far, this section has looked only at simplistic security, in that each issue has been taken separately. But this simple view of security breaks down when multiple, overlapping, or hierarchical VLANs are considered. If a user is a member of two different VLANs, then it must be decided which VLAN security policy takes precedence, or a policy-based system must be set up, which involves creating a third category of a dual-VLAN security policy containing both policies.

VLANs and ELANs

As mentioned before (see Chap. 4), the ATM Forum's LAN Emulation 1.0 is not considered a virtual LAN—since this section focuses on the transport of layer 3 protocols constituting a VLAN across an ATM switch backbone—but an Emulated LAN (ELAN). When ELANs and VLANs are connected in the same network, there can be conflicts between the two. For example, allocating QoS parameters to the virtual circuit(s) which constitute part of the virtual LAN as well as the synchronization of VC forwarding tables have implications for the integrity and stability of VLANs as well as network performance. If the switch becomes congested, then cells or cell streams representing layer 3 VLAN packets will be missed or dropped, and the VLAN will become unstable. When this happens, the non-ATM switched network is suffering "transit VLAN starvation" or, in the case of ATM networks, "VC starvation." One-to-one VLAN/ELAN mapping will resolve these issues.

In an ELAN, there is a strict priority order: if any cells are available for CBR (constant-bit-rate) traffic, they will be transmitted first. If no cells are available for CBR, then rtVBR (real-time variable-bit-rate) cells will be transmitted. If no cells are available for either CBR or rtVBR, then nrtVBR (nrt = non-real-time) cells will be transmitted, and so on until all service categories are examined. Since CBR traffic never produces a burst, having a strict priority is acceptable because there will always be bandwidth available for the other service categories. However, if VBR traffic is given a strict priority over other classes, then several VBR VCs may be starting to transmit at their PCR. The sum of these PCRs could exceed the available bandwidth, and the service categories with lower priority might be unable to get any cells out. This could cause timeouts in the higher-layer applications, such as a layer 3 VLAN, which would then have to retransmit the data contributing to even further congestion. The problems worsen if ABR and UBR are prioritized. Since the ABR and UBR service categories have been designed to utilize all the available bandwidth, the lower-priority service category will constantly be in starvation.

Enhancing VLAN Performance

Experienced users say that bandwidth utilization can drop by 40 to 60 percent with the installation of VLANs, and at a lower cost, too.

In a modern network, especially in the presence of an intranet where users do not know where the data resides, the old pattern of backbone congestion being limited by the 80/20 rule (20 percent inside the local area, 80 percent outside, handled by routers) has broken down, and with switches loading the traffic onto the backbone ever faster, VLANs in switched networks should be able to ameliorate the problems.

Switched VLANs with High-Speed Aggregates

VLANs must support a high-speed aggregate for a scalable network architecture. Layer 2 switches typically utilize an application-specific integrated circuit (ASIC) design, and this gives the switch its high speed and low cost (although ASIC design is expensive, as switches are now sold in such large numbers, the cost when spread across total expected sales is very small and allows such high integration that fewer processors are needed in a switch with ASICs). The switch is thus implemented primarily in silicon, that is, is hard-wired, unlike routers. The bridge group definitions are used to determine boundaries of the broadcast domain and maintain unique collision domains for each switch port. These implementations are fast and cheap, but not flexible. They do not adapt well to changing network requirements.

Network layer protocols are transparent to layer 2 devices, and IP subnet boundaries cannot be interpreted by standard switches. In fact, different workstations with different IP subnet addresses cannot communicate through the switches unless a routing function is introduced to resolve communication between different subnets. Since the distributed applications that warranted the introduction of switches are likely to be based on the TCP/IP protocol, if an edge router is used, then added latency has crept back in. This is where the layer 2 switch gains the advantage where resolving the addresses from different subnets.

A standard routing algorithm such as RIP (Routing Information Protocol) passes traffic between "route groups," but once the routes within a route group are learned, IP packets are forwarded through the network layer switching engine. The concept of "virtual subnet" arises from switched connection between different IP subnets. The multilayer switch has a combination of switching hardware and routing software. The main benefit over a router is speed, but additionally, because existing addressing schemes and connectivity preferences can be integrated into the switch configurations, a hierarchical network design can be supported. A disadvantage of defining VLANs at layer 3, however, where the packets can cross the backbone, can still be low performance. Inspecting layer 3 addresses in packets is more time-consuming than looking at MAC addresses in frames. For this reason, switches that use layer 3 information for VLAN definition are generally slower than those that use only layer 2 information. (It should be noted that this performance difference is true for most, but not all, vendor implementations.)

These concepts alone do not scale when the traffic pattern follows a more asymmetric model. Extending Ethernet to handle a traffic model where all communications converge on a single high-powered server would be like sending all the wastewater from hotel bedrooms to a single, same-size pipe. Backed-up bathrooms would result. This analogy is a good one since what is usually installed are "fat pipes"—a high-speed aggregation method. Typically, Ethernets are now aggregated into Fast Ethernets or ATM for a distributed backbone. In the future, Ethernet or Fast Ethernet LANs will be brought in by Gigabit Ethernet or, again, ATM. A clever development in early 1997 that makes the best use of existing technology aggregates Fast Ethernets by means of multiple Fast Ethernet pipes—needing almost as many ports out of the back of the switch as on the front. This is likely to be an interim solution to the problem, however, as the faster technologies arrive.

Now armed with a multilayer switch vendor of choice and a number of high-speed shared-media alternatives, design concepts of the virtual LAN can be extended to a more asymmetric traffic model. The integration of these new tools allows the virtual network model to scale and handle the bandwidth requirements generated when many clients converge on a single, large application server.

The VLAN and the WAN

VLANs can be extended across the WAN, although this will mean LAN broadcasts consuming WAN bandwidth. Although WAN bandwidth is expensive today, it is widely forecast to become much cheaper in the future. Routers, of course, filter broadcast traffic, and have for years been used to keep LAN broadcasts safely contained. However, even today, while still expensive, if WAN bandwidth is free or unlimited and underused for a particular organization, then extending layer 3 VLANs over a WAN should be considered in order to fulfill the ideals of flexibility discussed in Chap. 2. ELANs, of course, being ATM-based, can be extended across the WAN (see discussion below), and it is also possible for IP multicast groups functioning as VLANs to be extended across the WAN without wasting WAN bandwidth.

It is still not clear, as noted above, whether IP multicast groups should—or how they will—span WANs, or whether this proposed technology is intended for office interconnection or for remote single users coming into the LAN intermittently, but as it is related to backbone-spanning VLANs, the problems are similar. Again the router must be replaced by a routing function on a layer 3 switch if performance is to be good.

It is in the WAN that ATM with ELANs really comes into its own, at least as far as speed goes. Telcos are already offering a service to companies for LAN interconnection, achieved by means of ELANs invisibly to the customers. The QoS characteristics mean that priority traffic should reach its destination with the minimum of delay. However, within an ATM WAN, VCs traversing large hops often receive poor performance as a result of propagation delays, even when routers are replaced by ATM switches. This situation can be greatly improved if a higher priority is given to existing traffic, rather than to data entering the network, via a network access control scheme.

Remote VLAN Members

It is unclear how members of a VLAN in remote and branch offices can be included in a central VLAN. While remote logon to a central server via an authentication server is well understood, its logical equivalent still has to overcome several barriers. First, because management is central to virtual LANs, VLAN management capabilities must be extended outward to encompass the remote office. As a first stage, remote stackable hubs should be replaced with stackable switches with integral VLAN support for the remote office. This pushes VLANs out into the wide area where link speeds are a fraction of a fast LAN backbone or even PC-switch link and cost is at a premium. This will restrict the attractiveness of strategic VLANs.

A further problem is that, to date, all VLANs assume a continuous connection between the PC and the switch. To overcome this, switch vendors will need to develop the concepts of dynamic VLAN membership further to include intermittent connections and multiple VLAN memberships—for example, a salesperson in the sales VLAN will have one (static) address while sitting at the office and a different (dynamic) address while logging in remotely from a laptop. The implication for authentication servers is that they must accept both addresses as legitimate, but not both at the same time.

With remote users, the assignment of the VLAN IP address must be made after the user has been authenticated as a VLAN user on the central VLAN. This means that authentication access servers of all kinds must be extended to include tables of valid VLAN members. In the case of highly mobile users, PAP/CHAP authentication needs to be mapped to VLAN memberships.

Current Products Available

The VLAN Vendor Scene

Given the diversity of possible VLAN installations at present, it is wise to consider the major offerings, each on its own merit(s). In this section, therefore, the vendor scene is described with particular relevance to VLANs, and the major offerings are examined in detail. At the end of the chapter, you will find a section headed "Recommendations," which will aid in deciding on the merits of the various VLANs available.

The LAN and internetworking market is dominated by four giants: 3Com, Cisco, Bay Networks, and Cabletron. Newbridge Networks, now that it has acquired UB Networks, has risen to join them at over the magic billion-dollar-revenue mark. There are lesser vendors who also cannot be ignored, often because of their size in a different market, or because of their large installed base. These include Digital Equipment, Hewlett-Packard, and IBM. These vendors have often arisen from a base of competence in one area only to, nowadays, encompass all or most networking technologies. There are other companies which specialize in switching products, and which have made a name for themselves in that niche—these include FORE Systems, ATM purveyors, and Xylan (Ethernet switches). Most of these companies offer virtual LANs under one scheme or another. Other companies, such as Compaq, Crosscom, D-Link, and many others, do offer VLAN support on their switches but haven't yet made an impact on the VLAN scene, and are not covered here.

Many of the companies with something to say about VLANs, including summaries of their switching offerings (since switching and VLANs are, as we have seen, intrinsically tied together,) are briefly summarized below. Companies not covered here which do have VLAN offerings, such as LanOptics or Crosscom (now part of Olicom), are not covered because of their small market share in switching (the former company mentioned) or current lack of integration (the latter company has yet to integrate its two parts). In some cases, while vendors claim to have VLAN capabilities, in fact there is nothing yet to see except a simple port-switched VLAN; in other words, the switches can do it, but there is no management software to speak of.

In the summaries, the type of product is categorized by

- *Desktop*—for the small office or small group of users.

- *Workgroup*—for a group of users within an enterprise.

- *Departmental*—also known as *wiring closet,* these will often cover one whole floor of a building.

- *Enterprise*—these will aggregate all switched traffic within a large building.

- *Campus*—it is rare as yet for switches to aggregate campus traffic, but layer 3 devices are beginning to replace routers in this role.

There are no hard-and-fast rules as to which type of switch you need, and they can be mixed and matched to suit. The differences are not only the number of ports they can have but also the amount of memory for buffers and address tables and the speed of the processor.

There is also differentiation between chassis, standalone, and stackable switches. Again, for a large building, you will want a large chassis-based modular hub with cards for different functions that can be hot-swapped in and out; for a small office, a single, standalone switch may well do the job perfectly well. Stackables lie in between, as they can operate as stand-alones, but can also be conjoined to act as a larger switch. True stacks only look like one repeater to the network, but some are really cascading, in which case they quickly limit the distance the network can cover as they look like multiple repeaters to the management software.

Technology type is briefly mentioned in the switching tables, with the main type first (usually Ethernet) and any high-speed uplinks mentioned after (e.g., Ethernet, Fast Ethernet). A switch which is mainly an ATM device (i.e., has an ATM switching fabric) may have an Ethernet downlink, and so will be put down as "ATM, Ethernet."

After this brief look at all the major vendors in the field, there then follows detailed implementation notes on some of the most interesting offerings. There are examples of each standard used, plus any widely used proprietary VLANs. All the companies whose VLAN product are profiled in depth are either leaders of at least part of the LAN networking market or are well known for their VLAN, but the choice of vendor and VLAN scheme to highlight does not reflect any kind of ranking, nor does it indicate that VLAN schemes other than those chosen are not equally good. At the end of the chapter, you will find general comments on the VLANs available today.

Networking Vendors

3Com.   3Com's base of expertise was originally networking cards, and these still contribute a large percentage of the company's income and inform a lot of its corporate strategy, leading it to concentrate on the local area. However, by means of a series of acquisitions plus in-house development, 3Com has now become a leader in all areas of networking, although it is still best known for NICs, hubs, and LAN switches (see Table 9.1). There are switches in all LAN technology types, even the still-rare FDDI, which the company has offered for the last two years, but the most successful are Ethernet, Fast Ethernet, and ATM.

3Com's network management software, Transcend, is also becoming better known than erstwhile. Transcend VLANs are deployed with 3Com's broad range of high-function switches, boundary switches, hubs, routers, network interface cards, and network management appli-

TABLE 9-1

3Com Switching
Products

Product family	Switch product type	Technologies supported	Virtual LAN support
Superstack II	Wide range of stand-alone and stackable hubs and switches incorporating the former LinkSwitch and Superstack families	Ethernet/FDDI; Ethernet/Fast Ethernet; Ethernet/ATM; Ethernet/Ethernet, or token ring	Layer 2
CoreBuilder	Wide range of wiring-closet chassis hubs with switching functions including former LinkBuilder, LANPlex, and Cellplex families	One or all of Ethernet, Fast Ethernet, token ring, FDDI, ATM, and IP routing	Layers 2 and 3, VLANs, and ELANs

cations. 3Com has an emphasis on standards that runs across its entire product range, and its VLAN is, of course, strictly conformant to the emerging 802.1Q/1p standard. This scheme is covered in detail a little later, chosen for its strict conformance and the fact that it runs on all 3Com switches, including ATM, and incorporates everything into the Transcend management system. In addition, 3Com is one of the few vendors to offer policy-based VLANs. Each switch can support up to 16 VLANs. However, there can be more than 16 VLANs in the entire network by connecting the 16 switched VLANs to other VLANs using a router or by using Chassis products. 3Com's VLANs work with its Net-Builder router range.

Network management is a key element within 3Com's three-phase VLAN strategy: Transcend Virtual Networking (TVN).

1. Transcend VLANs simplify network moves and changes and improve server access. Most phase 1 functionality is shipping today.

2. Transcend VLANs will enable customers to reduce the use of LAN routers, simplify switch configuration, and introduce standards-based multivendor interoperability. Phase 2 functionality is being delivered during 1997.

3. VLANs and VLAN membership are no longer static or semistatic designations, but are dynamic, with the virtual structure of the network responding in accordance with demand for services. Phase 3 functionality began shipping in May 1997.

3Com is on course to complete the final phases of Transcend VLANs in 1997, and has delivered on all incremental steps to date. Transcend VLANs include VLANs and other enabling technologies and features to build virtual networks, such as LAN Emulation (LANE), network multimedia support, distributed network monitoring and analysis, and advanced automated network management. A detailed explanation of 3Com's VLANs follows later.

Agile Networks. In October 1996, Lucent Technologies Inc. (the split-off technology company from AT&T) acquired datanetworking switch manufacturer Agile Networks Inc. (see list of products in Table 9.2), marking Lucent's first move toward offering its own datanetworking technology. Lucent had been reselling hubs, switches, and routers from Bay Networks Inc., which it will continue to do so, but has not yet mapped out specific plans for developing and integrating products and technology acquired from Agile, leaving network managers little information with which to weigh Lucent's potential as a strategic partner in data technologies. Agile, acquired for an undisclosed sum, brings only one product type under Lucent's umbrella, but Agile's expertise with routing software may offer longer-term value. Currently, Agile has used its expertise to create extensive VLAN capabilities available—the switch automatically learns end stations' addresses and places them in VLANs. As with other vendors, however, the technique for Ethernet will remain proprietary for the foreseeable future. LANE, under ATM, is standard.

Agile's VLAN capability is heavily weighted toward ELAN integration and support, and does offer some policy-based filtering. Agile's VLAN manager is called ATMman, and it features the following:

■ *Global discovery of end stations.* By communication with the ATMizer switches, it finds out all layer 1, 2, and 3 protocols and VLAN memberships.

TABLE 9-2

Agile Networks Switching Products

Product family	Product type	Technologies supported	Virtual LAN support
121 ATMizer Ethernet Switch	Standalone workgroup	Ethernet, ATM	Layers 2 and 3
125 ATMizer ATM and Ethernet Switch	Standalone enterprise	ATM, Ethernet	Layers 2 and 3, ELANs

- *Layer 3 name services.* For IP end stations, ATMman queries the network DNS, NIS, or IP host name. For IPX, decode the SAP broadcasts.

- *Event logging of moves, adds, and changes.* Captures, displays, and logs all end-station changes on the network.

- *Policy management of moves, adds, and changes.* Follows administrator-set restrictions for security and control purposes.

- *IP address management.* Administrators are informed of any changed or duplicated IP addresses that appear.

- *End-station locator.* By specifying a host name or MAC or IP address, ATMman identifies the segment to which any end station is attached and any VLAN memberships it may own.

- *ATM link topology map.* ATM link status is displayed in color.

- *Mappings between VLANs and physical LANs.* Indicates which end stations are on any physical segment and which segments have memberships of any VLAN.

- *VLAN partitioning.* Manual creation or partitioning of any VLAN.

- *ATM and Ethernet Statistics.* Traffic and error statistics with graphical display.

- *SNMP manager with secure sets—and acknowledged traps.*

ATMman runs on a UNIX workstation and is integrated with HP OpenView, and the database is object-oriented.

Bay Networks. Bay Networks was originally formed from a merger of Wellfleet Communications and SynOptics and covers all major sectors of the LAN market with routers, hubs, and switches (see list of Bay Networks products in Table 9.3). It is also successful in network management software, with its Optivity platform. Bay Networks has also been actively acquiring companies in the last few years, notably for VLANs Centillion, with its ATM switches.

In the Centillion or the 5000BH card, which is a Centillion switch, a mechanism defines a proprietary VLAN or ELAN. There are a couple of limitations with the ELANs: (1) it is possible to configure only one token-ring ELAN (soon to be 32), and source route bridging is not supported with LAN Emulation (it is supported with the GIGArray proprietary mode, and will be supported in a next software upgrade). The Centillion will get layer 3 VLANs toward the end of 1997. This is purely port switching—it is possible to define a port as part of an IP subnet or

TABLE 9-3

Bay Networks
Switching Products

Product family	Product type	Technologies supported	Virtual LAN support
BayStack 301	Standalone workgroup	Ethernet, Fast Ethernet	Layer 2
Centillion 100	Chassis-based enterprise	Ethernet ATM	LANE
Ethernet Workgroup Switch	Standalone	Ethernet	Layer 3
Switch Node	Chassis-based enterprise	Ethernet, Fast Ethernet; soon to have ATM, FDDI, Gigabit Ethernet	Layers 2 and 3
LattisSwitch 28000 family	Chassis-based enterprise	Ethernet, Fast Ethernet	Layer 3

layer 2 (MAC) subnet. The newest member of the switching family, the proprietary Switch Node, will run a protocol-sensitive 802.1Q VLAN between the switches in a frame-based environment. Bay here creates virtual ports and splits the pipe, multiplexing the signals along it. Bay Networks is lagging behind its rivals with VLANs, although plans are well advanced for a graphical manager with drag-and-drop facilities. This will be in Bay Networks' management platform, Optivity, which is a good element manager but has not yet advanced to the point of Cabletron's Spectrum for integrated management or 3Com's Transcend for overall management of VLANs. At present, however, each device has to be physically configured separately to create or change VLANs. Within Optivity, end stations will autolearn their IP subnet (although for IPX, manual configuration is needed). A control center, due out in mid-1997, will allow VLANs to be built on a workstation, which will then configure the devices (routers, hubs, LAN switches, ATM switches). A division is to be made between configuring the VLANs and configuring the devices attached to the network. ELANs and 802.1Q VLANs will interoperate. At present, Bay uses 802.10 between routers, and will not be building VLANs to this standard.

Cabletron Systems. Until recently, Cabletron (see product list in Table 9.4) stood out against the trend and did not acquire companies, nor did it expand from its original base of hubs and switches only, prefer-

TABLE 9-4

Cabletron Systems
Switching Products

Product family	Product type	Technologies supported	Virtual LAN support
SmartSwitch Ethernet	Chassis-based enterprise	Ethernet	Layers 2 and 3
SmartSwitch FDDI	Chassis-based enterprise	FDDI	Layers 2 and 3
SmartSwitch 2200	Standalone desktop, workgroup	Ethernet, Fast Ethernet with optional FDDI or ATM	Layers 2 and 3
7C03/4/7 SmartSwitch	Desktop, workgroup	Ethernet, FDDI, ATM	Layers 2 and 3
8H02-16 SmartSwitch	Standalone workgroup	Ethernet	Layers 2 and 3
SmartSwitch 6000	Chassis-based, departmental	Ethernet, Fast Ethernet with FDDI or ATM connectivity	Layers 2 and 3
6-Slot MMAC Plus	Chassis-based departmental, enterprise	Ethernet, token ring, Fast Ethernet, FDDI, ATM	Layers 2 and 3
MMAC Plus	Chassis-based enterprise	Ethernet, token ring, Fast Ethernet, FDDI, ATM	Layers 2 and 3
3C05 ATX LAN Switch	Modular standalone	Ethernet, token ring, Fast Ethernet, FDDI	Layer 3
ZX	Desktop	ATM	ELAN
SFCS-200 (from FORE)	Enterprise, workgroup	ATM	ELAN
SFCS-1000 (from FORE)	Enterprise	ATM	ELAN
CTM	Enterprise, carrier class	ATM	ELAN

ring to resell routing expertise. Recently, however, the company has given in to the trend and purchased smaller companies to augment its product range. In the realm of switching, the only acquisition has been the switching business of SMC, manufacturers of adapter cards and switches.

Cabletron has a far wider range of switches than listed in Table 9.4, but the table includes all the switches that the company puts forward as important to its VLAN strategy; lack of space precludes the inclusion of more. Most of the rest of the range comprises switch modules for the chassis models. By the end of 1997, Cabletron is expected to bring out a small, desktop switch resulting from the SMC purchase. The company is also hoping to standardize the names of the switches on the SmartSwitch brand, including the well-liked and well-known MMAC+— this is to distance them from the older MMAC hub range. Cabletron has one of the most advanced VLANs around, SecureFAST. One of the first of the block, this is still the most flexible and is beginning to be widely installed. Cabletron's current VLAN scheme is based on broadcast control and full link utilization, but the newest version implements policies. Cabletron has one of the most advanced VLANs available today. Not only does it already have policy-based filtering, the company is also attempting to grapple with the idea of interworking with Directories, and is negotiating with Novell and Microsoft to try to integrate its VLAN database with their directory services. Cabletron's policy server will be able to interrogate the directory server. The 1998 version is expected to have Novell login by VLAN. Also, Cabletron plans to implement SecureID from Security Dynamics on its VLAN by 1998.

Cabletron's network management system, Spectrum, has gained much success and many plaudits in the last year as a platform to rival Hewlett-Packard's. So far, however, it has not been fully integrated into Cabletron's SecureFAST virtual networking architecture. Cabletron's VLAN scheme is widely praised and is one of the most advanced in terms of software, the interface, and policies. A detailed explanation of Cabletron's VLANs follows later.

Cisco Systems.   Cisco is the largest of the networking-only vendors in terms of annual revenue and shipments (see products list in Table 9.5). It comes from the world of routing, and its routers still have the largest market share of any company. (Figures vary, but, for example, Cisco claims that 80 percent of the Internet runs on its routers.) It has now broadened out into switching by means of several major acquisitions. Cisco's weakness has been in its network management offerings, but this may be

TABLE 9-5

Cisco Systems
Switching Products

Product family	Product type	Technologies supported	Virtual LAN support
Catalyst 1200	Standalone work-group switch	Ethernet, FDDI	Layers 2 and 3
Catalyst 1600	Rack-based enterprise switch	Token ring, ATM (LANE), FDDI	Layer 2
Catalyst 2100	Standalone desktop and workgroup switch	Ethernet, Fast Ethernet	Layer 2
Catalyst 2800	Standalone desktop and workgroup switch	Ethernet, FDDI	Layer 2
Catalyst 2900	Standalone enterprise and workgroup switch	Ethernet, Fast Ethernet	Layer 2 and 3
Catalyst 3000	Stackable and standalone enterprise and workgroup	Ethernet, ATM, 100VG AnyLAN	Layer 2
Catalyst 5000	Chassis-based switching hub for enterprise and workgroup	Ethernet, ATM, FDDI	Layer 2 and 3
LightStream	Enterprise and workgroup	ATM	ELANs

changing now that it has an alliance with Hewlett-Packard, a major force in network management systems with its OpenView platform.

Cisco's switches have come from various acquisitions over the years, beginning with Crescendo and adding Kalpana, pioneer in Ethernet switching, LightStream's ATM switches, Grand Junction (Fast Ethernet), Granite Systems (Gigabit Ethernet), and Nashoba Networks (token ring), and Stratacom (WAN switches and carrier-class ATM switches). This disparity has not aided integration, and the VLANs may not work across all the models.

Cisco's VLAN technology is an extension of its CiscoFusion architecture, which attempts to unify switching, ATM, and routing. Cisco began working on its VLAN technology in 1994 but did not lay out its five-phase strategy for it until 1996:

Phase 1. Filtering tables controlling broadcasts between switching points and user stations and reducing the number of broadcasts flowing through switches. Phase 1 was implemented in 1994.

Phase 2. In 1995 Cisco completed this phase with the release of the 7000 line of routers and the Catalyst 500 modular switch, which introduced packet tagging.

Phase 3. To address campuswide VLAN switched internetworks across all major backbone types (FDDI, Fast Ethernet, and ATM), Cisco is currently developing a mapping protocol, VLAN Trunk Protocol (VTP), in conjunction with packet tagging technology. VTP will allow network administrators to manage distributed VLAN topologies with a central administration point. VTP will enable mapping to future 802.1Q standards.

Phase 4. Automated configuration working together with VTP.

Phase 5. Introduction of additional applications and security options for VLANs. Network administrators will be able to define and track user access rights and define policies to balance networking services.

Cisco has initiated partnerships with Intel Corp. and Xpoint Technologies to extend VLAN functions to network interface cards and plans to support VLANs up to the application layer in 1997.

Cisco has chosen to develop its VLAN on the "other" standard, 802.10, although there are plans to add a 802.1Q VLAN at a later date. Cisco has had to build in mapping to the 802.1Q standard, which at present is not released. Cisco has further difficulties because of its loyalty to the cause of routing, which means that it must continue to develop its IOS suite, which, although it runs across both switches and routers today, is far more attuned to the routing world. Compatibility with IOS is a must for Cisco, given these needs. (A detailed explanation of Cisco's VLANs follows later.)

Digital Equipment. Best known as a systems vendor, Digital also has a networking business unit with a wide range of equipment, including hubs, switches, and routers (see products list in Table 9.6). This contributes only a small part of Digital's total revenue, but is one of the more successful parts of the company. At the time of writing, there was a rumor that Digital was about to sell off its networking products division.

Digital has an innovative range of hubs and switches that can work standalone or stacked, or be clipped into a wall-mounted rack of its own design. This means that there cannot be easy interoperability and none of this range will mount into a normal rack with other vendors' equipment, but it does mean that the same piece of equipment can be used in a variety of situations as a company's needs change. This factor does not apply to the GIGAswitch range. All can be managed by EnVisn, the management platform, which also incorporates the virtual LAN software.

TABLE 9-6

Digital Equipment
Switching Products

Product family	Product type	Technologies supported	Virtual LAN support
DECswitch 400 LAN	Chassis- and rack-based, departmental	Ethernet, ATM	Layers 2 and 3
DECswitch 900 EE	Chassis- and rack-based or standalone enterprise	Ethernet	None
DECswitch 900 EF/FO	Chassis- and rack-based or standalone enterprise	Ethernet, FDDI	Yes, proprietary
GIGAswitch/ ATM	Chassis- and rack-based or standalone enterprise	ATM	LANE
GIGAswitch/ FDDI	Chassis- and rack-based or standalone enterprise	ATM, FDDI	Yes, proprietary
ATMSwitch 900	Chassis or standalone workgroup	ATM, Ethernet	LANE, layer 2
PEswitch 900TX	Chassis- and rack-based or standalone workgroup	Ethernet, FDDI	Yes, proprietary
Portswitch 900	Chassis-based or standalone workgroup	Ethernet	None
MultiSwitch 600/300	Chassis-based enterprise and workgroup	Ethernet, Fast Ethernet, token ring, FDDI, ATM	Layers 2 and 3
VNswitch EX	Chassis-based enterprise	Fast Ethernet	Layers 2 and 3
VNswitch EF	Chassis-based enterprise	FDDI	Layers 2 and 3
VNswitch EA	Chassis-based enterprise	ATM	LANE

The MultiSwitch is Digital's newest product, and the 600 is aimed at workgroups while the 300 is for departments; apart from greater density, they are the same. The VNswitch family has up to three VNbuses (each running at 400 Mbits/s) on the backplane, with the possibility of running VLANs across all them. It is on this platform that Digital's VLANs are most developed, although, as seen above, VLANs can run across most of the switches. There is a limit of 32 VLANs per module and 64 per VNbus. These VLANs can then be extended to the ATM cloud via the VNswitch EA ATM switch. VLANs can also go down to a DEChub 900 or Multiswitch 900, allowing VLANs to go right across an enterprise. Although Digital is very proud of its VLAN capabilities, some of its switches do not support any virtual networking, and others use a purely proprietary scheme. Digital has developed its VLANs to the nascent

802.1Q standard, but also offers optional add-ons to enhance performance. Digitals VLANs offer policies, unlike most schemes offered today, but they are as yet no more than a set of filters. A detailed explanation of Digital's VLANs follows later.

FORE Systems.   FORE started out as the leading ATM company for LANs, and this continues to be its specialty, although from its acquisition of Alantec, it also has a range of Ethernet and Fast Ethernet switches (see products list in Table 9.7).

Given FORE's concentration on ATM, it is natural that LANE should be its speciality in virtual networking, and therefore its scheme is concerned with ELANs. FORE's overall network management software is called *ForeView*, which includes a virtual LAN application that simplifies the configuration of ELANs by enabling users to perform drag-and-drop operations to graphically specify separate ELANs and the membership of hosts within them. (A more detailed explanation of FORE's ELANs follows later.)

Hewlett-Packard.   HP is known as a systems vendor, but is also best known in networking as the purveyor of the highly successful network management platform OpenView. In addition to this, it also has a range of stackable hubs and switches. HP was the main developer of 100VG-AnyLAN, which it continues to support alongside 100Base-T. Hewlett-

TABLE 9-7

FORE Systems
Switching Products

Product family	Product type	Technologies supported	Virtual LAN support
PowerHub 4000	Chassis-based workgroup	Ethernet, Fast Ethernet, ATM, FDDI	Layers 2 and 3
PowerHub 6000	Chassis-based workgroup	Ethernet, Fast Ethernet, ATM, FDDI	Layers 2 and 3
PowerHub 7000	Chassis-based enterprise	Ethernet, Fast Ethernet, ATM, FDDI	Layers 2 and 3
ASX200WG	Workgroup	ATM, Ethernet	Yes, via LANE
ASX200BX	Backbone	ATM, Ethernet	Yes, via LANE
ASX-1000	Backbone	ATM, Ethernet	Yes, via LANE

TABLE 9-8

Hewlett-Packard
Switching Products

Product family	Product type	Technologies supported	Virtual LAN support
AdvanceStack 10/100 LAN Switch	Standalone workgroup	Ethernet, Fast Ethernet, 100VG-AnyLAN	None
AdvanceStack Switch 200	Standalone workgroup	Ethernet, 100VG-AnyLAN	None
AdvanceStack Switch 2000	Chassis- and rack-based or standalone workgroup, backbone	Ethernet, Fast Ethernet, 100VG-AnyLAN	Layer 2
AdvanceStack 100	Standalone workgroup	Ethernet, FDDI	No

Packard plays only in the workgroup sector of the switching market. (See products list in Table 9.8.)

Although justifiably proud of their range of switches, HP has a long way to go to catch up in the VLAN game—most of its switches as yet have no support for them and there is no support for layer 3 at all.

IBM.   IBM is, of course, also best known as a systems vendor, but it has a wide range of networking equipment centering on MAUs and switches (see products list in Table 9.9), although it also has its own routing technology. The networking side of the business gain contributes only a small part of the company's revenues.

Today, IBM's LAN switches include the 8271 EtherStreamer and 8272 LANStreamer switches, which are used to segment LANs. In the future, IBM will create a broad switch family suitable for workgroup and campus connections that fully integrates with IBM's ATM switches. These switches will connect to ATM or fast, switched-LAN networks via fat pipes and will support combinations of shared and dedicated connections. The planned products include standalone versions and stackable and modular, intelligent hub blades. IBM's switches support adaptive cut-through and store-and-forward operations. For enhanced management and performance, IBM will offer virtual LAN and broadcast management support.

In October 1996, IBM announced the Nways Manager allowing the configuration, control, and maintenance of LANs, virtual LANs, ATM

TABLE 9-9

IBM Switching
Products

Product family	Product type	Technologies supported	Virtual LAN support
8271 EtherStreamer	Rack-based or standalone workgroup	Ethernet	Layers 2 and 3
8272 Nways Token-Ring LANSwitch	Rack-based or standalone workgroup	Token ring	Layers 2 and 3
8285 Nways ATM Workgroup Switch	Rack-based or standalone workgroup	ATM	Yes, proprietary

networks, and WANs. The announcement included new versions of the existing Nways Manager for Windows, Nways Campus Manager LAN, ATM and ReMon for AIX, Nways Campus Manager Suite for AIX, and two new management applications: Nways RouteSwitch Network Manager and Nways RouteTracker Manager. IBM also announced a new Nways Manager for Enterprise Networks, combining the functions of the Nways BroadBand Switch Manager, the Router and Bridge Manager, and the SNA Alert Manager into a single product called the *Nways Enterprise Manager.* This included VLAN management using a drag-and-drop interface, graphical device management, and policy-based virtual LAN management. The New Nways Enterprise Manager, in conjunction with the other Nways managers, allows end-to-end management of an entire network, combining the functions of the Nways BroadBand Switch Manager, the router and Bridge Manager, and the SNA Alert Manager. IBM VLANs can now be created in mixed environments of Ethernet, token ring, or ATM Emulated LAN (ELAN) segments.

Intel. This is a recently created division of Intel, with a fascinating history. First, in 1990, there was Dataco, a medium-sized (for those days) Danish LAN company. Shortly thereafter it was bought up by the British company Cray Communications (nothing to do with Cray Research, the supercomputer manufacturer). Then, in 1996, after a lot of financial difficulty, Cray Communications changed and split into Anite Systems, an integrator; and Case Technology, which continued to manufacture. The original Dataco was still producing the LAN equipment, now including switches, in Denmark as part of Case Technology. The financial problems were not, however, fully solved by this move, and

TABLE 9-10

Intel Switching
Products

Product family	Product type	Technologies supported	Virtual LAN support
Matchbox switch	Standalone backbone	Ethernet, Fast Ethernet, ATM	Layers 2 and 3

early in 1997, Case Technology sold off the Danish LAN division to Intel. Intel (see switching product in Table 9.10) had been making a move into networking for some time with the purchase of Thomas-Conrad, adding token-ring adapter cards and hubs; and of Networth, for Ethernet hubs. Intel, of course, already produces Ethernet and Fast Ethernet adapter cards. Case Technology has now joined this family.

As Cray Communications, the company was very keen on VLANs very early. Indeed, one of the reasons for its struggle was probably its accent on technology rather than successful marketing, and the technologists there could see the need for VLANs in any fully switched network, which was, after all, what the company was selling. In the push to survive, development has not continued, and although the original innovative scheme has rather taken a back seat, it is better than many others as it is one of the few policy-based VLANs, and it will no doubt be brought to the forefront again once the standards are set.

Madge Networks. UK company Madge Networks always concentrated on selling into the IBM-dominated token-ring market (with most of its sales in the United States), but 2 years ago, Madge bought the Israeli company LANNET, respected maker of high-speed Ethernet hubs and switches. The ranges have been amalgamated and Madge now sells a full range of networking equipment (see Table 9.11). Before this, Madge followed IBM into 25-Mbit/s ATM, which has not, as yet anyway, really taken off. This shows a lack of any input into VLAN development.

The LANSwitch is from the LANNET range; RingSwitch shows its original Madge ancestry with its emphasis on token ring. The 3LS switch, released in January 1997, was the world's first all-silicon multilayer IP/IPX LAN switch. The company also at the same time revealed its strategy for multilayer IP/IPX switching across Ethernet, Fast Ethernet, token-ring, FDDI, and ATM networks. The first product of this strategy was the 3LS layer 3 switching module for Madge's LANswitch modular switching hub.

TABLE 9-11

Madge Networks
Switching Products

Product family	Product type	Technologies supported	Virtual LAN support
LANswitch	Chassis-based enterprise	Ethernet, Fast Ethernet, token ring, FDDI	Layer 2
Smart RingSwitch	Rack-based enterprise	Token ring, ATM, FDDI	Layers 2 and 3
3LS MultiLayer Switch	Rack-based enterprise	Ethernet, Fast Ethernet, FDDI, ATM	Layers 2 and 3

With the addition of the 3LS module, the LANswitch hub becomes a multilayer IP/IPX switch, capable of switching both IP and IPX packets at either layer 2 or layer 3—at throughput rates of up to 1 million packets per second (mps) while executing up to 4000 user-configurable filters across Ethernet, Fast Ethernet, FDDI, and ATM. The 3LS module for the Madge LANswitch hub supports multilayer IP/IPX switching with an aggregate throughput of 1.28 Gbits/s in a single chassis and delivers full wire speed over a range of physical LAN infrastructures including Ethernet, Fast Ethernet, FDDI, and ATM. Two or more 3LS modules can be installed in the LANswitch chassis to provide an exceptionally high degree of fault tolerance. If a 3LS module fails, the IP/IPX switching load is transferred to the backup module transparently to network users. The LANswitch 3LS plugs directly into the Cellenium Bus, which has a backplane speed of 2.56 Gbits/s, giving it a multigigabit switching-speed routing throughput.

On the LANNET side, developments have been hampered by the cultural change involved for LANNET staff—many left the conjoined company, and Madge has suffered from a lack of Ethernet expertise. Both Madge and LANNET were developing ATM products before the takeover. Now, the company has yet to fully integrate the different sides of the business and has no overall element management system. Because of all these factors, Madge has not yet released a fully functioning VLAN product. However, on the LANNET side, developments did begin and there are two variants—10Base-FL or 10Base-T—each of which takes up one slot in the Ethernet chassis. Each switch port can be used to define a VLAN with parallel connections across hubs for high throughput. MultiMan is LANNET's network management system (NMS) that runs on 386-based personal computers, Suns SPARCstations, AIX, or UNIX platforms. It is capable of managing not only LANNET hubs but also RND bridge/routers and any other SNMP-manageable elements in the network. MultiMan works with all the leading manager-of-manager NMSs: HP OpenView, Sun Solstice SunNet Manager, and IBM SystemView NetView/6000. Per port performance of Ethernet modules is available, displaying extensive traffic statistics. There are, within MultiMan or available as optional add-ons, several specialist programs. Virtual-Master is the part that can be used to set up workgroups across geographically dispersed locations. It is with this software that the beginnings of true workgroup virtual LANs of users on ports anywhere on the network can be seen. Virtual LANs can be set up and torn down at will. Users may belong to concentric workgroups.

Newbridge Networks.   Newbridge designs, manufactures, and markets managed solutions for WANs and corporate LANs (see products list in Table 9.12). Traditionally, it has been a WAN vendor but has expanded into the LAN arena. The WAN history shows its product range: high-capacity ATM multiplexers for use in corporate networks and telephone company central offices down to low-capacity, narrowband feeder multiplexers in addition to the range of switches shown here. Early in 1997, Newbridge acquired UB Networks, with which it has long had an association but this is still too new for full integration of products. Newbridge has not released a specialized VLAN plan although one has been in development, but it is expected to use UB networks advanced scheme in the future.

The former UB Networks, originally known as *Ungermann-Bass* and until recently a subsidiary of Tandem the mainframe vendor, was bought by Newbridge Networks in 1997 and is in the process of becoming part of the VIVID group in the Entr@Networking Division. Its area of expertise was always in hubs, but over the last 2 years it has developed its own switching technology, with which it is having success. It is on the GeoLAN/500 and GeoSwitch/155 chassis hubs and switches that UB's VLANs have been implemented, with trunking allowed between the 500 and 100. The GeoSwitch/155 is, however, in the process of being phased out, to be replaced by the VIVID workgroup ATM switch. The GeoLan/100 is the well-known Acess/One hub, renamed. UB will have two types of VLAN available, one proprietary and one based on 802.1Q. The present VLAN is 802.10-based, and UB was aligned with Cisco on this. All this is now changed, and its current status is uncertain. The proprietary scheme is up and running; the other is being built now. A detailed explanation of the VLANs can be found in the following paragraphs.

TABLE 9-12

Newbridge Networks Switching Products

Product family	Product type	Technologies supported	Virtual LAN support
VIVID Workgroup Switch	Rack-based or standalone enterprise and workgroup	Ethernet, ATM	Layers 2 and 3
VIVID Yellow Ridge	Rack-based or standalone workgroup	Ethernet, ATM	Layers 2 and 3
VIVID Blue Ridge	Rack-based or standalone workgroup	Token ring, ATM	Layers 2 and 3
VIVID ATM Workgroup Switch	Rack-based or standalone enterprise and workgroup	ATM	LANE
GeoLan/100	Chassis hub with switching modules	Ethernet, token ring, FDDI, LocalTalk	Layer 2
GeoRim/E	Workgroup switch	Ethernet with uplinks to FDDI, Fast Ethernet, ATM	Layer 2
GeoLAN/500	Chassis hub and switch with switching modules	ATM basis, modules for any technology can be added, but no VLANs with token ring	Layer 2

Optical Data Systems (ODS). ODS VLAN-related products are listed in Table 9.13.

ODS has several other models, but they are all based around the Warrior switch, and, in the strategy rolled out by ODS in January 1995, it is the Infinite Switching product line that will form ODS's first VLANs. ODS admits that this area of virtual switching holds the most promise as well as the most challenges for the implementer. Infinite Switching can be configured through the command line or the GUI interface. The same functionality is available with either interface. The command-line interface can be accessed in band by an ASCII terminal or modem attached directly to the EIA/TIA-232-C interface on the Virtual Network Controller module. In addition, remove out-of-band access using a Telnet session is supported through any LAN or ATM interface.

TABLE 9-13

ODS Switching
Products

Product family	Product type	Technologies supported	Virtual LAN support
Warrior 1	Chassis-based workgroup	Ethernet; Fast Ethernet, ATM on some models	Layers 2 and 3
Warrior 2060	Standalone enterprise	Ethernet, token ring, Fast Ethernet, ATM, FDDI	Layers 2 and 3
Warrior 1 2060	Standalone workgroup	Ethernet, token ring, Fast Ethernet, ATM, FDDI	Layers 2 and 3
Infinity Desktop Switch	Chassis-based desktop	Ethernet, Fast Ethernet on some models	Layers 2 and 3
Massively Parallel ATM/OC3 Switch	Chassis-based enterprise	ATM, Ethernet	LANE
Massively Parallel ATM25 Switch	Chassis-based enterprise	Ethernet, 25-Mbit/s ATM	Layers 2 and 3
Massively Parallel Ethernet Switch	Chassis-based enterprise	Ethernet, ATM	Layers 2 and 3

ODS Infinity LanVision offers tools for device management, fault management, configuration management, performance management, security management, and other functions. The Infinite Switching strategy integrates two ODS applications: ODSVision and ODS VLANVision.

Plaintree Systems. Founded in 1988, Plaintree Systems is not a new start-up, but is still one of the smaller companies in this part of the market. It is headquartered in Massachusetts (USA) and its research, development, and manufacturing facilities are in Ontario (Canada). Plaintree Systems is a member of the Gigabit Ethernet Alliance, the 100VG-AnyLAN Forum, and the ATM Forum. It has been an enthusiastic adopter of VLANs. (See products list in Table 9.14.)

The WaveSwitch 9200 Gigabit Ethernet Switch is one of the first operational gigabit switches, with powerful networking features grouped under the name *NextWave Switching*. For example, it increases performance by controlling broadcast and multicast traffic and simplifies network operation by automatically configuring network parameters. Wave-View is Plaintree's intuitive GUI network management software, but at

TABLE 9-14

Plaintree Systems
Switching Products

Product family	Product type	Technologies supported	Virtual LAN support
WaveSwitch 100	Chassis- and rack-based or standalone enterprise and workgroup	Ethernet, 100VG-AnyLAN	None
WaveSwitch 4 + 1/4 + 4	Standalone workgroup	Ethernet, Fast Ethernet	None
WaveSwitch 1018/9	Standalone workgroup	Ethernet, Fast Ethernet	Layers 2 and 3
WaveSwitch 1216	Standalone	Ethernet, Fast Ethernet	Layers 2 and 3
WaveSwitch 4205	Rack-mounted or standalone workgroup	Fast Ethernet	Layer 2
WaveSwitch 4800	Rack-based enterprise	Ethernet, Fast Ethernet, FDDI, 100VG-AnyLAN	Layers 2 and 3
WaveSwitch 9200		Gigabit Ethernet, FDDI Fast Ethernet, Ethernet	

present the VLAN functions are not fully integrated into it. Plaintree VLANs implement a selectable multilayer, multilevel architecture. VLAN features include automatic membership learning and tracking across the enterprise, autoconfigurability, and improved eavesdropping protection. The VLAN processing is performed in hardware.

Whittaker Xyplex. Whittaker's VLAN-related products are listed in Table 9.15.

Xyplex was best known for its routing chassis hubs, to which switching functions were later added. It has been through a checkered history in recent years—bought by Raytheon in 1996, it was not developed further but was sold on again in 1997 to Whittaker. This is a good match, since Whittaker is known for its ATM switches and will no doubt integrate the Xyplex range in time. However, it is still too soon to see real results in this direction. At Xyplex, there had been substantial investment in VLANs, with a scheme that encompassed both layers 2 and 3 (MAC-address-based and IP-based); however, the management and inter-

Product family	Product type	Technologies supported	Virtual LAN support
1700	Standalone workgroup	Ethernet, Fast Ethernet, ATM	Layer 2
708	Chassis-based enterprise and workgroup	Ethernet, ATM	Layer 2
Network 9000	Chassis-based enterprise	Ethernet, FDDI, ATM	Layers 2 and 3
SX 6600	Standalone workgroup	Ethernet	None

face is still not highly developed, and it remains to be seen how future developments will go under the Whittaker management.

Xylan. Xylan is a relatively new company, offering LAN switching from its foundation in 1993. It became interested in the possibilities of VLANs right away and now sells its products (see list in Table 9.16) with the aid of its well-developed scheme. Because it entered the arena early, it used the 802.10 standard, like Cisco, for its tagging.

Xylan offers PizzaSwitch (so named because it is no larger than a pizza box), OmniSwitch, and OmniVision Network Management Suite, which includes SWITCH Manager and an interface for configuring AutoTracker virtual LANs. Users can select any ports on one or more OmniSwitches and group them into a VLAN. Users can configure up to

Product family	Product type	Technologies supported	Virtual LAN support
OmniSwitch	Chassis-based enterprise	Ethernet, Fast Ethernet, 100VG-AnyLAN, FDDI, ATM	Layers 2 and 3
PizzaSwitch	Standalone workgroup	Ethernet, Fast Ethernet, 100VG-AnyLAN, FDDI, ATM	Layers 2 and 3

65,000 VLANs in a network and can link VLANs with the routing functions incorporated into the OmniSwitch or through external routers. Xylan's VLAN scheme is much praised and remains one of the few to offer policies. A more detailed explanation of Xylan's VLAN follows.

The *OmniSwitch* is a flexible, powerful, and highly reliable switching platform for campus applications. It combines an innovative hardware architecture with a sophisticated feature set, yet it's so inexpensive that it can serve as a basic network building block. OmniSwitch is uniquely versatile. It switches any combination of Ethernet, token ring, FDDI, TP-PMD, Fast Ethernet, and ATM at wire speed with automatic any-to-any translation, over twisted-pair, coaxial, or fiberoptic cable. It routes IP and IPX. It connects to network segments, file servers, or individual workstations. OmniCell is the advanced ATM switching capability to be added to the OmniSwitch in 1997. It will coexist with the current frame switching capability, and the two will be linked via a high-speed SAR process. An OmniSwitch with OmniCell will support up to 64 OC-3 ports, up to 16 OC-12 ports, and up to 128 25M ports. OmniCell uses a powerful new fabric architecture that is optimized for support of a mix of CBR, VBR, ABR, and UBR traffic, in very bursty applications.

PizzaSwitch is a family of low-cost, midrange LAN switching products from Xylan. The PizzaSwitch is the most powerful, versatile LAN switch of its size. It switches locally among Ethernet segments and devices, and links them at high speed to servers and backbones using FDDI, Fast Ethernet, and ATM.

Detailed Implementation Notes

3Com

3Com views VLANs as an intrinsic part of IP switching and end-to-end connectivity; the real value of broadcast control is seen only when it *is* end-to-end. 3Com has always been supportive of open standards, and it is fitting, therefore, that its VLAN is based firmly on the 802.1Q protocol and 802.1p GARP (GARP allows complex layer 3 switching without the need for a router; any edge device can request to join a multicast)—of course, it also helps to drive these nascent standards forward. Although there is an emphasis on IP; in fact, any suitable protocol can be at the heart of the network, such as IGMP (Internet Group Management Protocol) snooping.

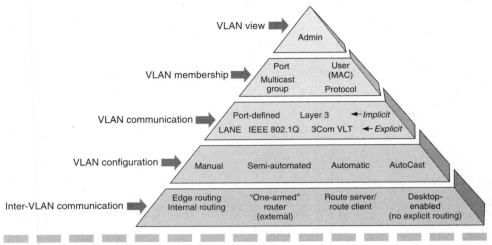

VLAN view ➡

Admin

VLAN membership ➡

Port   User (MAC)

Multicast group

Protocol

VLAN communication ➡

Port-defined   Layer 3   ◄ *Implicit*

LANE   IEEE 802.1Q   3Com VLT   ◄ *Explicit*

VLAN configuration ➡

Manual   Semi-automated   Automatic   AutoCast

Inter-VLAN communication ➡

Edge routing Internal routing   "One-armed" router (external)   Route server/ route client   Desktop-enabled (no explicit routing)

Figure 9-1   3Com VLAN structure. (*Source: 3Com*)

3Com has a three-part Transcend Networking framework intended to make networks easy to use, as well as simpler and more cost-effective to design and manage. Transcend Networking framework includes hardware and software solutions for (as 3Com's slogan goes): scaling the performance, extending the reach, and managing the growth of enterprise networks. Transcend is also the name of 3Com's network management system, into which the VLAN structure falls. 3Com Transcend Enterprise Manager (TEM) software provides complete management of 3Com's network systems, including configuration-change management, fault-alarm management, and performance management as well as VLAN management. In Fig. 9.1, it is clear that management, which sits at the apex of the pyramid, with all further functionality dependent on it, is the primary driver.

3Com has extended VLAN management capability within TEM software across all its switching products. Available on both Windows and UNIX, TEM manages VLANs on 3Com's switches (including ATM) and NETBuilder II routers. Transcend Enterprise Manager UNIX 4.2 also features distributed management capabilities via HP OpenView 4.1. TEM Windows 6.0 is integrated with HP OpenView for Windows 7.2d. Additional features of TEM Windows 6.0 on OpenView include the ability to receive a page for alarms and compatibility with any Winsock-compliant TCP/IP stack. StatusWatch is a value-added capability within Transcend Enterprise Manager that correlates and aggregates switch status.

VLAN Membership Definition. 3Com will support its own Virtual LAN Trunking (VLT) frame-tagging method in many of its products until the 802.1Q standard is finalized (and thereafter will continue to offer added value). LANE is supported in all of 3Com's ATM switching products (see descriptions in the following paragraphs). Transcend VLANs enable definition of VLANs by all methods:

- Switch port group
- MAC address
- Network layer information (including by protocol type and/or IP address)
- Multicast group
- Policy-based VLANs

With 3Com VLANs defined by MAC-layer address, initial configuration, as well as moves and changes, can be automated. Transcend VLAN Manager also enables the network administrator manually to change a user's VLAN membership when necessary, such as when an employee transfers from one department to another. Port-switching and MAC-address VLANs are available today, as part of phase 1 of 3Com's strategy.

For customers with large numbers of IP users, 3Com's CoreBuilder switches support VLANs defined by IP address. Normally, IP requires the network administrator to physically go to the user's workstation and reconfigure that user's IP address after a move. For better flexibility, 3Com allows multiple "virtual" IP subnets to coexist on a single physical segment, a technique called *multinetting,* which enables administrators to maintain logical groups of users without the constraint of physical location. Many customers may wish to replicate the multiprotocol broadcast domains of their existing router-based infrastructures. For this purpose, CoreBuilder switches also support defining VLANs by protocol (such as IPX, DECnet, NetBIOS). Network layer VLANs are now available as phase 2 of 3Com's VLAN strategy.

Multicasts. Transcend VLANs offers two methods for distributed control of multicasts without heavy reliance on routing. The first method enables the switch itself to define "AutoCast VLANs" on the basis of multicast groups. The technique used in switch-based AutoCast VLANs is called *Internet Group Management Protocol* (IGMP) snooping. It operates by having the switch observe user-initiated requests to belong to a particular IP multicast group (e.g., a video-based training session). The switch then dynamically defines IP multicast groups according to those requests, forwarding the multicast traffic only to those ports with participating users and blocking it on all other ports.

The second method goes one step further in distributing the intelligence necessary for multicast control by allowing the NIC driver in the user's workstation to control the multicast filters of the switch port to which it is attached. The driver initiates a message from the NIC to the switch, telling the switch whether to forward a particular multicast on that port. This method has advantages over IGMP snooping in that it supports all multicast traffic (not just IP multicasts), and it uses the processing power in the workstation CPU, enabling deployment of simple workgroup switches. However, this method is optimized for an architecture based on a single user per switch port and requires upgraded NIC drivers. 3Com will offer both methods of multicast control in order to best meet diverse customer needs. Multicasting is part of 3Com's phase 2, to be delivered later in 1997.

A further future development is in the network management station—it is to become a Java-enabled, push-technology device, the "Java Virtual Machine," which will extend the existing boxes' capabilities.

Policy-Based VLANs. In the final phase (3) of the migration to VLANs, the Transcend VLAN architecture enables the organization to reach an extremely high level of automation in the administration of the network. Transcend VLANs will enable the network to dynamically self-configure on the basis of policies set by the network administrator, and on the particular applications and/or network services that are accessed by each user at a given time. The policy-based VLAN administration functionality will be available starting with TEM 4.2 UNIX. This new functionality is based on next-generation SmartAgents (first deployed in the CELLplex [now CoreBuilder]) that can enforce network administration policies. These administration policies could apply to network configuration (such as auto or dynamic VLANs based on some logic), distributed network monitoring, and distributed event handling. These next-generation (Java based) SmartAgents could also serve as proxy agents for other devices, and so these smarter agents need not be deployed in every device—only the core switches LDAP will be supported.

To simplify initial assignment, 3Com's VLAN Policy Services offer several policy options to assign users to VLANs; MAC address, IP subnet, and protocol type. To ease moves, adds, and changes, if a user moves within the network, 3Com switches will sniff the user's MAC address and will automatically assign the user to the correct VLAN, stored in a master map in the VLAN server within Transcend Enterprise Manager. This eliminates the need for IT staff to intercede and manually reconfigure users into their VLANs.

VLAN Policy Services can increase network security by automatically placing unrecognized network users into a default VLAN, secure from the rest of the network. For example, when any user plugs into any switch port, the SmartAgent software on the switch checks the user's MAC address against the VLAN map within Transcend Enterprise Manager's VLAN server. If the user has an unknown MAC address, the switch will alert the network management station and place the user in a secure VLAN.

3Com's VLAN Policy Services work by way of software resident in several products. Transcend Enterprise Manager includes comprehensive VLAN management tools, VLAN policy-setting software, and one or more VLAN servers that hold the global mapping database. Second, SmartAgent software embedded in 3Com switches enforce the centrally set policies. Transcend Enterprise Manager downloads policies directly into each switch's SmartAgent software. When users connect to switch ports, the SmartAgent software determines their VLAN memberships according to the enforced policy and by consulting with a VLAN server, and autoconfigures each user into the correct VLAN. This globally distributed SmartAgent software-based architecture ensures that the solution can scale to the largest networks.

VLAN Configuration. 3Com has developed an intuitive, graphical VLAN management platform, Transcend VLAN Manager. The VLAN architecture supports all three levels of automation for VLAN configuration: manual, semiautomated, and fully automated.

1. *The Default VLAN and moving ports from the default VLAN.* On each switch, VLAN 1 is the Default VLAN of the switch; it has two properties:
- It contains all the ports on a new or initialized switch.
- It is the only VLAN which allows an SNMP network manager to access the management agent of the unit.

By default, if a device attached to a port should belong to a VLAN other than the Default VLAN, you need to use the VLAN Setup screen to place the port in that VLAN.

2. *Connecting VLANs to a router.* If the devices in a VLAN need to talk to devices in a different VLAN, each VLAN requires a connection to a router. Communication between VLANs can take place only if they are all connected to the router. A VLAN not connected to a router is isolated. VLANs are typically connected to routers using backbone ports. Backbone ports have the following attributes:

- Addresses received on backbone ports are not stored in the switch database.
- Frames with unknown addresses are forwarded to backbone ports.

If a switch is connected to a router using backbone ports, it is necessary to specify one backbone port for each VLAN connected to the router. For connection to other switches, it is also necessary to use backbone ports. In addition, to make the switch-to-switch connections more cost-effective, it is possible to specify that one port forms part of a *Virtual LAN Trunk* (VLT), a connection which carries traffic for multiple VLANs between switch units. If both ends of a switch-to-switch connection are connected as part of a VLT, only that one connection is needed for all the VLANs. VLTs cannot be used for switch-router links. If a backbone port on one VLAN is specified as part of a VLT, that backbone port will become a backbone port for all the VLANs on the switch—even if they had no backbone port before. If you subsequently disable the VLT function on that port, the port becomes the backbone port for the Default VLAN (VLAN 1) and all other VLANs lose their backbone ports. VLT uses a single spanning tree to control VLAN traffic. While this limits performance and scalability, it is all that is built into the first version of 802.1Q A VLT 802.1Q gateway will be released. While 802.1Q supports 12-byte VLAN definitions, VLT at present uses only 4, allowing a maximum of 16 VLANs rather than 802.1Q's 4096. If more are needed, then an ATM mesh with LANE needs to be implemented—that gives unlimited VLANs and ELANs. 3Com's VLANs and ELANS can run across all of its switches.

3. *Using AutoSelect VLAN mode.* By default, all ports on the switch use Port VLAN mode, where each switch port is manually placed in the required VLAN. The switch allows some ports to use another mode, AutoSelect VLAN mode. In this mode, the ports are automatically placed in the required VLAN by referring to a VLAN server database in 3Com's Transcend Enterprise Manager v6.0 for Windows. AutoSelect VLAN mode works as follows:

- When an end station is connected to a switch or moves from one port to another, the switch learns the MAC address of the end station.
- If the relevant port uses AutoSelect VLAN mode, the switch

    refers to the VLAN server database which contains a VLAN allocation for each MAC address.

- Having obtained the VLAN allocation for the MAC address, the switch places the relevant port in that VLAN.

AutoSelect VLAN mode has an advantage over Port VLAN mode because once the VLAN Server database is set up correctly, end stations can be moved to other ports or other switch units and the VLAN allocation of each end station is automatically reconfigured. In AutoSelect VLAN mode, note the following:

- Specific IP address and community strings are needed for the VLAN server.
- VLAN 15 cannot be used.
- If a port has been configured as a backbone port or as part of a VLT, the port cannot use AutoSelect VLAN mode.
- If a port has a permanent address stored against it in the switch database, the port cannot use AutoSelect VLAN mode.

3Com recommends that each switch port is connected to a single end station. To connect a port to multiple end stations, it must be specified that the port uses Port VLAN mode.

4. *Using nonroutable protocols.* If nonroutable protocols are running on the network, devices within one VLAN will not be able to communicate with devices in a different VLAN unless bridged by a CoreBuilder.

5. *Using unique MAC addresses.* In the case of a server with multiple network adapters to the switch, 3Com recommends that each network adapter have a unique MAC address.

6. *Setting up VLANs on the switch.* The VLAN Setup screen allows

- Assignment of ports to VLANs, if those ports use Port VLAN mode
- Definition of a backbone port for each VLAN
- A view of the VLAN setup information for the switch

To access the VLAN Setup screen: (*a*) from the VT100 Main Menu, select SWITCH MANAGEMENT—the Switch Management screen appears; (*b*) in the Management Level field, choose VLAN; then (*c*) choose the SETUP button—the VLAN Setup screen appears as shown in Fig. 9.2.

The screen shows the following:

- Port ID 1,2,3,...,24,25,26 (3C16900A) 1,2,3,...,12,13,14 (3C16901A). This field allows entry of the ID of the port that you want to set up.
- VLAN ID 1,2,3,...,16. If the port specified in the Port ID field

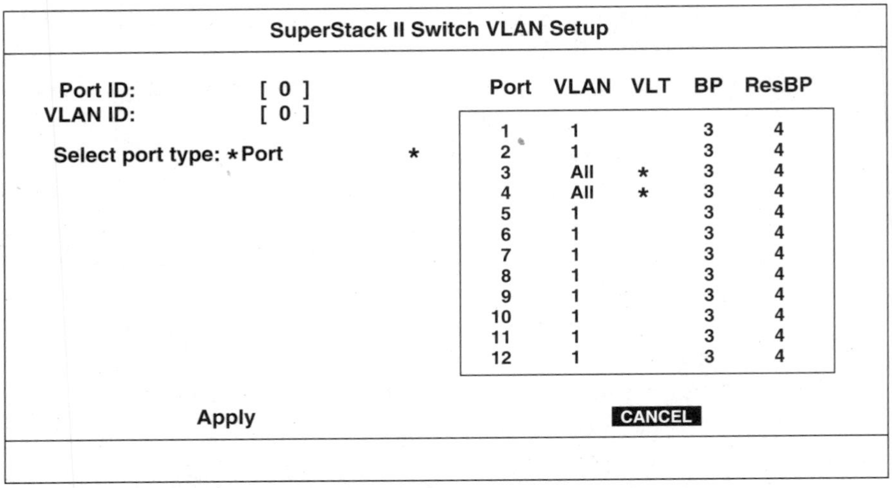

SuperStack II Switch VLAN Setup					

Port ID: [0]
VLAN ID: [0]

Select port type: *Port   *

Port	VLAN	VLT	BP	ResBP
1	1		3	4
2	1		3	4
3	All	*	3	4
4	All	*	3	4
5	1		3	4
6	1		3	4
7	1		3	4
8	1		3	4
9	1		3	4
10	1		3	4
11	1		3	4
12	1		3	4

Apply   CANCEL

Figure 9-2   3Com VLAN Setup screen. (*Source: 3Com*)

uses Port VLAN mode, this field allows you to enter the ID of the VLAN to which the port is to be assigned. If the port uses AutoSelect VLAN mode, the VLAN ID cannot be defined. By default, all ports use Port VLAN mode and belong to the Default VLAN (VLAN 1).

If using AutoSelect VLAN mode, VLAN 15 cannot be used. Also, if using the spanning-tree protocol, VLAN 16 cannot be used.

Next, select Port Type Port/Backbone Port. If the port specified in the Port ID field uses Port VLAN mode, this field allows you to specify whether the port is a backbone port unless 802.1D mode is selected. If the port uses AutoSelect VLAN mode, you cannot specify that it is a backbone port. A backbone port is used to connect each VLAN to the backbone of your network, and has the following attributes:

- Addresses received on the port are not stored in the switch database.
- Frames with unknown addresses received by the switch are forwarded to the port.
- Any port in a VLAN can be designated as the backbone port for that VLAN, but there can be only one backbone port per VLAN. By default, all ports belong to the Default VLAN (VLAN 1); because of this, an unconfigured switch unit can have only one backbone port.

If a plug-in module is fitted into a switch, this automatically becomes the backbone port for the Default VLAN when on powerup or initialization. If a switch has no plug-in module, but does have a transceiver module, this becomes the backbone port for the Default VLAN on powerup or initialization. A listbox contains the following fields:

- *Port*—the port ID for the entry.
- *VLAN*—the ID of the VLAN to which the port belongs.
- *VLT*—shows if the port forms part of a *Virtual LAN Trunk* (VLT), a connection which carries traffic for multiple VLANs between Switch units.
- *BP*—the backbone port for the VLAN specified in the VLAN field.
- *ResBP*—this field displays the resilient backbone port for the VLAN, if one exists.
- *APPLY*—this button applies any changes to the VLAN database.

7. *Assigning a port to a VLAN when using Port VLAN mode*

- In the Port ID field, type the ID of the required port.
- In the VLAN ID field, type the ID of the required VLAN.
- Select APPLY.

Initially, all switch ports belong to the Default VLAN (VLAN 1). In early implementations, this was the only VLAN which allowed an SNMP network manager to access the management agent. If all ports were removed from VLAN 1, then an SNMP network manager could not manage the switch. This problem has now been solved.

8. *Specifying a backbone port.*   To specify a backbone port: (*a*) in the Port ID field, type the ID of the required port; (*b*) in the VLAN ID field, type the ID of the required VLAN; (*c*) in the Select Port Type field, select Backbone Port; then (*d*) select APPLY.

9. *Setting up VLANs using AutoSelect VLAN mode.*   To set up VLANs using AutoSelect VLAN mode, you need to

- Specify information about the VLAN server
- Specify that the switch unit, or individual ports on the unit, use AutoSelect VLAN mode
- Specify information about the VLAN server

The VLAN Server screen allows specification of information about the VLAN server; to access this screen: (*a*) from the VT100 Main Menu, select SWITCH MANAGEMENT. The Switch Management screen appears; (*b*) in the Management Level field, choose VLAN; then (*c*) choose the SERVER button. The screen shows the following:

- *VLAN Server IP Address:* Enter the IP address of your VLAN server in this field.
- *Backup VLAN Server IP Address.* This field allows you to enter the IP address of a backup VLAN server, which can be used to supply VLAN allocations when the switch cannot access the main VLAN server.
- *VLAN Server Community String.* This field allows you to enter a community string for the VLAN server(s). The default community string is public.
- *Throttle 0,...,99999.* This field allows specification of the time delay, in milliseconds, between the transmission of VLAN allocation requests to the server. The time delay is used to avoid placing an excessive workload on the VLAN server. The default setting for this field is 50 ms.
- *Poll Period.* This read-only field shows the time interval, in seconds, between successive polls of the VLAN server. The switch polls the VLAN server once every poll period to check for any changes.

Inter-VLAN Communication. 3Com plans to support the following types of inter-VLAN communication, because each can have a place, depending on the customer's overall network environment:

- Edge routing
- The "one-armed" router
- MPOA
- Desktop-enabled zero-hop routing
- Servers on multiple VLANs

3Com is presently delivering a solution based on the edge-routing model, integrating the routing function into its CoreBuilder switches. While routing will be the primary method for inter-VLAN communication for some time, it is not the only method. Transcend VLANs also enable end stations (usually servers) to be members of more than one VLAN, effectively providing an application layer gateway between VLANs.

As the Transcend VLANs solution moves forward, membership in a given VLAN will become less a static designation and more a dynamic one. This membership can be governed either by the switch or, eventually, by the desktop/NIC driver. As VLANs become more dynamic, the need for routing inter-VLAN traffic will disappear; if an application calls for two or more end stations to communicate for a period of time, they are simply placed in the same VLAN (the switch-governed model), or they join the same VLAN (the desktop/NIC driver-governed model) for the required period of time.

LANE. Also, 3Com's Transcend Enterprise Manager (TEM) provides LANE visualization and management tools. Using TEM for LANE, network managers can

- Automatically discover and display the LANE service infrastructure along with the ATM physical network infrastructure
- Automatically discover and map Emulated LANs
- Map LANE client-server relationships and proxy LEC to port associations
- Perform virtual circuit tracing between LANE elements (segment to segment, LEC to LEC, or LEC to LES/BUS) and also map physical paths over the ATM infrastructure
- Graphically display LEC and LES/BUS performance statistics
- Manage the LEEKS database synchronization
- Manage LANE service redundancy and automatic failure mechanism
- Isolate LANE service faults and correlate effected devices and segments

The LANE standard as implemented in the CoreBuilder 7000 family allows devices to communicate whether attached to Ethernet, Fast Ethernet, token ring, or ATM links.

Cabletron Systems

Cabletron coined the word *Synthesis* as the name of the architecture of its virtual LAN and networking strategies because it synthesizes the functions of a multiprotocol router into the network without the limitations posed by routers. Synthesis encompasses three principal areas:

- *Infrastructure*—the physical layer of the network, which includes hubs, LAN and ATM switches, SNA/LAN integration products, routers, and Cabletron's SecureFast Virtual Network Services (VNS).
- *Automated management*—SecureFast VNS-supported applications (see below under SecureFast Virtual Network).
- *Support services*—assists users in migrating to Cabletron's Synthesis

through Cabletron's global support services in over 100 strategic locations provided by a support staff of over 1600. These services include customer training, seminars, technical support, and service.

Cabletron's VLAN and Vnet strategy addresses the need for users to

- Construct simple layer 2 VLANs based on switch port or MAC address assignments
- Move to layer 3 VLANs to add greater flexibility and functionality
- Implement switched virtual networks using the existing LAN infrastructure, precluding the need to migrate to ATM to achieve their objectives

Cabletron says its vision of virtual LAN and networking technology is predicated on its commitment to

- Protect the investment of its customers
- Substantially reduce the cost of network administration
- Provide a migration path to multiprotocol, connection-oriented, virtual networks that support any LAN topology
- Preclude the need for users to wait for ATM standards and to migrate to ATM

To do this, Cabletron has a two-phase VLAN and networking strategy:

Phase 1: layer 2 VLANs. Phase 1 of Cabletron's VLAN strategy lets users partition networks into VLAN domains on the basis of either port-switching or MAC addresses. VLAN mapping to switch ports or MAC addresses is configured by a VLAN management application. The switches do a peer-to-peer distribution of VLAN mappings. Also under phase 1, Cabletron's VLAN strategy adds network layer (layer 3) flexibility and functionality to layer 2 VLANs. It provides multiprotocol support and virtual routing to let users communicate between VLANs. With VLAN policy management, Cabletron users can elect to switch or route packets between VLANs. Other vendors provide routing only between VLANs. Cabletron's layer 2 switching is performed on source and destination address pairs. All other vendors perform switching on the destination address only, which does not provide the granularity of control and management. Phase 1 of Cabletron's VLAN strategy is available now.

Phase 2: SecureFast Virtual Network. Phase 2 of Cabletron's solution to virtual technology is its virtual networking strategy, called the *Secure-*

Fast Virtual Network—a virtual networking strategy that combines Cabletron's SecureFast Switching architecture with its VNS application to produce a SecureFast Virtual Network. VNS is a software application that distributes conventional routing functionality throughout a network:

■ *Connection-oriented network*—sets up connections between source and destination end stations by examining the address header of only the first packet in a stream, and passes packets over the established path, guaranteeing the delivery of packets to their correct destinations. This also delivers fast data transfer.

■ *Multiprotocol and multimedia support*—provides multiprotocol support for all networking topologies: Ethernet, token ring, FDDI, and ATM LANE.

■ *Policy Management Service*—processes calls using policy-based management to control access and priorities to the network to guarantee network security. This is done once at call setup time. It is administered and enforced from a central location for security.

■ *Call Accounting Service*—maintains call statistics for each call to support capacity planning and/or customer billing to recover third-party costs.

■ Call Management Service—automatically establishes types of calls: dynamically switched or administratively defined permanent virtual circuits over point-to-point, point-to-multipoint, and multipoint to multipoint circuits. It enables network administrators to observe active calls and probe them using a protocol analyzer via the GUI.

■ *Virtual Routing Service*—established through policy-based management and can be defined by an endpoint address range or a collection of selected endpoints. Workgroups are hierarchical, which means that one or more workgroups can be contained in another workgroup. Policies attached to a workgroup remain in force regardless of where the endpoint appears on the network.

The SecureFast Virtual Network has three main components:

■ *SecureFast Packet Switches (SFPSs)*—perform bridging, switching, and routing for Ethernet, token-ring, FDDI, and SNA packets with the performance of ATM switches; handle both packets and ATM cells for migration to ATM. Cabletron supports its SFS technology across all of its product families to provide customer investment protection and a migration path to ATM switching technology.

■ *A SecureFast Virtual Network Server* (SFVNS)—provides virtual routing, policy management, call management, and call accounting services.

■ *The SecureFast Virtual Network Management Application*—provides configuration management for virtual routing, policy definition, call management configuration, and call accounting.

SecureFast Architecture.   The SecureFast architecture is based around Cabletron's SmartSwitches but can also run on any vendor's equipment that can run the code (in C++). Although Cabletron is planning to go to 802.1Q later in 1997, at present SecureFast uses VLSP (Virtual Link State Protocol), a proprietary OSPF (Open Shortest Path First)-based connection. It allows the support of

■ Fully active mesh topologies

■ Best end-to-end path determination

■ Call distribution over equal-cost paths

■ Automatic rerouting on network link or switch failure

The main difference between VLSP and OSPF is that OSPF routers are based around IP addresses; VLSP is based on Cabletron SecureFast switches and interswitch links, with each switch represented by its MAC address. VLSP works with a single flat area, thus allowing every switch to have a full view of the topology. A gateway will be available and, for those customers who already have the current VLAN, a replacement will be offered. Cabletron chose this method to enable connection orientation. There is no reason why any protocol should not be used, for example, Ipsilon's IP Switching or Cisco's Tag Switching, if Cabletron customers want this.

VLAN Membership Definition.   Cabletron defines packets assigned to specific VLANs via standard addressing schemes inherent to the specific LAN topology. If a packet arrives untagged, it goes to VLAN 0. In the future, the NIC will tag the frame, as allowed for in 802.1Q. Today, the application is port-based and frames get tagged by the switch but, unlike 802.1Q, the inbound frames are not tagged, only outbound. For multicasts, it is essential for the NIC to tag the frames—this has to be addressed.

Users can be in up to eight VLANs and, when VLAN wizards (a VLAN "wizard" is an autoregistration tool—a sort of smart agent) arrive, up to eight of those, too. And these can be closed (i.e., packets remain local, but members in different VLANs can communicate) or open (where all inter-VLAN communication must go through a router) VLANs. As Cabletron's policies are brought in, later in 1997, then the router can be replaced by a switch without losing security.

VLAN Manager.   Cabletron's VLAN manager is a Spectra-server-based engine integrated into Spectrum, Cabletron's network management system. It also runs on several other SNMP-based management platforms, including HP's OpenView, Sun's Solstice SunNet Manager, and IBM's NetView/6000. The VLAN manager tool allows configuration and monitoring of the VLAN through a graphical interface.

At the time of writing, Cabletron had to run two servers on one computer, but was almost able to have both functions in one server sharing the same code. A Web-based VLAN manager is also about to be released, built on Java.

To configure the workstation or PC, the VLAN manager needs to know

- The IP address (or, if the IP address is constantly changing, as with DHCP, then the MAC address)
- Subnet mask
- Default gateway

Today's SecureFast supports MAC, IP, and IPX. The next version will also include AppleTalk, DECnet, and NetBIOS. Where IP is used, it is likely that a router will connect the VLANs, as in 802.1Q. Cabletron recommends a gradual migration to a fully switched network.

When the switches are connected together for the first time, if they see only one MAC address, they assume that it is an interswitch link and self-configure all the relevant technologies. If a router is on the net, the switches use RIP and forward the packets. If OSPF, the switch floods its VLAN—the router belongs to all VLANs and picks it up. The VLAN manager builds a Delta (Cabletron term) of a user table and network topology and then updates all the switches. Because each switch knows its own VLAN data, VLAN manager can be taken off its system, but it will be needed in the case of moves, adds, or changes. If a default gateway is set to itself, it thinks it knows the whole network, and will not keep an ARP cache, and thus will not send out ARP broadcasts (which can impede performance). If the switch receives an ARP, it looks at it and, if it knows which station sent it, swallows it. In this way, the switch acts like a router. If it does not know the sender, it looks at its two neighbors (each switch knows only its neighbors), and then issues an ARP

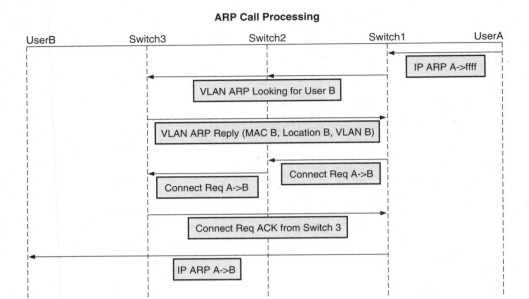

Figure 9-3 Cabletron ARP call processing. (*Source: Cabletron*)

request. In this way there is no broadcasting unless the user is not found. The switch recognizes and sends the packet back with information, using OSPF to pick the route (assuming permissions). If a second switch sends an ARP, the first switch sends a receipt back. Figure 9.3 shows how all this works.

SAP, like ARP, is another noisy protocol—and only other SAP users need to see it. Here, the virtual domain recognizes SAP broadcasts and tells the other switches to send SAPs only to those end stations which issue SAPs. It is a good idea to put all SAP users into a secure VLAN together.

Policy Server.   Where every user has a secure ID, a policy server acts as an access list, as on a router. For example

<username> is in X,Y,Z VLANs.

VLAN Z can talk to VLAN Y.

In X this user can log in 8:30 A.M. to 10:00 P.M. Monday through Friday. For VLAN setup, too, policies can be set, as such

VLAN A can talk to VLAN B between 9:00 A.M. and 6:00 P.M. via IP and IPX.

All policies are stored on the central policy server, which will in time also be the accounting server, so that <username> gets the data bill, and in turn this is linked to a purchase ledger database. This will probably not be put to such a purpose, but may well be useful for bandwidth-use policies. For example, in a campus with several companies sharing an ATM backbone, it would be possible to set up PVCs and charge for traffic use. This is also a way for telcos to implement Virtual Private Networks, and ADSL and video-on-demand services will employ this method. It can all be built onto Cabletron's Virtual Network Server (VNS), a Sun SparcStation in a specialist box. Today, this is only possible on FDDI, however.

VLAN Manager Interface.   There are a number of ways to access data in the database, all shown by using the Delta table. Figure 9.4 shows one such view.

Figure 9-4
Cabletron's Secure-
Fast VLAN Manager.
(*Source: Cabletron*)

To start up a VLAN, simply click on the CREATE VLAN button on the database toolbar and give it a name. From the users in the database, drag and drop from the database to the new VLAN (or to any other VLAN). Another view shows which users are on which switches, arranged hierarchically. A topology view allows managers to see the location of the switches and any of VLAN's members. A closed VLAN is indicated by a key.

The Connection table shows

- *Who is talking to whom*
- *Provisioning*—the physical PVCs are set here
- *Application preference*—IP address, user name, host name, etc.

The Directory is at the heart of the database. It contains all the users, including their locations, what network type they are on (IP, NetBIOS, etc.), and, last but not least, who they are by name and title.

The VLANdetail table gives the information on who is in which VLAN and also offers security options (this feeds the Delta table the keys).

Cisco Systems

Cisco regards VLANs as an integral part of its CiscoFusion architecture, which gives a strategy for the integration of multilayer switching throughout campus networks. Cisco has never sold hubs but embraced switching at the earliest opportunity instead. A Cisco spokesperson has said that by the year 2000, there will be no more shared ports and all networks will be one port per user. Given this belief, it is therefore to Cisco's advantage to make its switched networks as efficient and manageable as possible. Cisco has a roadmap for its VLAN development extending over 5 years, most of which has now been delivered, which goes a long way toward fulfilling these aims:

Phase 1: broadcast-controlled. These reduce broadcast traffic and improve performance by means of filtering tables that keep details of ports and MAC addresses. This degrades switch performance due to added latency. It is also not scalable. This type of VLAN was introduced in 1994.

Phase 2: bandwidth-managed. VLANs could be exchanged between LAN switches and routers by means of packet tagging (based on 801.10), allowing VLANs to scale to the campus and cross Fast Ethernet and FDDI hubs and switches, and also allow load distribution. This type of VLAN became available in 1995.

Phase 3: campuswide. As well as the campuswide tagging already discussed, Cisco introduced a VLAN Trunk Protocol (VTP) that maps and configures VLANs across the campus irrespective of the LAN technologies in use. This has been made an integral part of Cisco's well-known *Internetworking Operating System* (IOS), which runs on all of its routers and some of its switches. VTP is both a switch-to-switch and switch-to-router management protocol that manages all changes to VLANs and replicates all VLAN changes across the network. VTP was released in 1996.

Phase 4: change-managed. The network becomes self-configuring with initial VLAN assignment based on MAC address, but subsequent amendments are made according to network address or other data. In this phase, also, a series of integrated configuration components will be introduced: VCS (Virtual Configuration Server) will supply information to Ethernet, Fast Ethernet, and FDDI networks while a LECS (LAN Emulation Configuration Server) will serve clients on ATM switches. A centralized database of all VLAN/ELAN members will also be possible. This is all due to be completed by the end of 1997.

Phase 5: policy-driven. For policies to be effective, Cisco will provide tighter integration to layer 7 (application) of the OSI reference model for integration with group applications such as Lotus Notes and Novell NetWare. Cisco is in fact already working with Microsoft to integrate its ActiveX—this will enable NT Server Directory Services to integrate with Cisco VLANs. This is due to come to fruition by the end of 1998.

Unlike many other vendors, Cisco has developed its VLAN on the 802.10 headers, not 802.1Q. This is because it originally wanted to give access to the FDDI backbones for its VLANs, and the 801.10 tags provided a field in which a VLAN tag could be inserted and defined a process for packet fragmentation and reassembly that was consistent with FDDI specifications. For Fast Ethernet, where there is no packet fragmentation, Cisco developed its own InterSwitch Link (ISL) protocol that enables over 1000 VLANs to be set up without requiring any fragmentation and reassembly. Thus, while all VLANs are proprietary at present, Cisco's is even less near standard than most. The company has, however, promised to migrate to 802.1Q at a later date. It has worked with the 802.1Q committee to incorporate the best of its own scheme to the 802.1Q standard.

Because Cisco is not tied to 802.1Q, it can implement one spanning tree per VLAN without having to apologize for being nonstandard. This gives significant performance improvement.

VlanDirector. VlanDirector is an SNMP-based switch management application that provides graphics-based VLAN management capabilities for Cisco switches. Integral components of VlanDirector include graphical mapping utilities for viewing and configuring logically defined workgroups, drag-and-drop port-level configuration options for assigning users to VLANs, automated link assignment settings for managing VLANs campuswide, integration with common SNMP management platforms for consolidating system resources, and detailed reporting functions for maintaining audit trails.

VlanDirector features are summarized here:

- Provides auto topology discovery and screen displays of Cisco interconnected switches and routers within the campus

- Represents a logical overlay on top of a physical network for VLAN design, configuration, and monitoring

- Integrates with CiscoView for drag-and-drop graphical configuration management

- Delivers an easy method for adding, deleting, and modifying VLANs within the campus

- Provides comprehensive VLAN configuration support for Catalyst switches

- Verifies data so that users can accurately configure VLANs across the campus

- Provides cutaway screen displays of configured VLANs, including switches, links, and ports

- Offers discrepancy reporting of conflicting VLAN configurations

- Provides name search functions for locating and monitoring specific VLANs

- Displays operational status of ports per VLAN

- Includes link management utilities for adding and deleting VLANs across interswitch VLAN links

- Integrates with SNMP management platforms, including Solstice SunNet Manager and HP OpenView

Configuring VLANs across the Campus. VlanDirector lets users configure their VLANs by first discovering and mapping the physical topology of the campus using a reliable discovery process. Logical VLAN topologies can then be superimposed on top of the physical network and dynamically linked on the basis of a real-time connectivity model.

VlanDirector accomplishes this task automatically, requiring minimal intervention by the network manager. Network managers can fine-tune the configuration of their VLANs across the campus by changing the VlanDirector link configuration preference options. These options include shortest-path calculations, the number of redundant paths required, and link availability relative to VLANs already assigned. VlanDirector guides users through the option selections and then displays the resultant settings.

Configuration Consistency. VlanDirector offers a utility for maintaining VLAN configuration consistency across internetworked Cisco switches and routers. This is an integral part of the discovery process and is continuously updated by VlanDirector. When a configuration mismatch is discovered, VlanDirector automatically displays a window that flags the error and indicates the source of the problem.

Tracking VLANs. VlanDirector offers a series of graphical views for checking, auditing, and verifying configured VLANs within a network. These views are created using simple pull-down commands. These views can be filtered according to the VLAN or group of VLANs selected. They are reproduced as detailed printed reports that can be used to generate audit trails and track VLAN configurations.

At the core of VlanDirector is a drag-and-drop configuration interface for assigning switch ports to VLAN names, a feature provided by tight integration between CiscoView, Cisco's device management application (see the following paragraphs), and VlanDirector's user assignment windows. This level of integration simplifies the process of assigning users to a VLAN, and greatly reduces the skill level required when adding new users to a VLAN. CiscoView is dynamically launched from VlanDirector with simple mouse-click operations. CiscoView clearly identifies the VLAN configuration of every port displayed with color representations and legends, providing powerful monitoring functions and eliminating any undocumented configuration settings.

Constructing VLANs. VlanDirector offers a simple configuration utility for adding, modifying, and deleting VLANs to a campus network of Catalyst switches. This utility is part of the main VlanDirector window and is used to create and view every VLAN in the network. Network managers can highlight any of the VLAN names in the VlanDirector window and quickly pull up a VLAN topology or display a view that shows all other switch ports that belong to the selected VLAN. (See Fig. 9.5.)

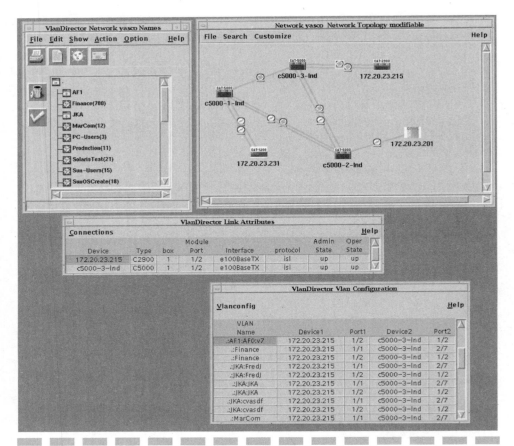

Figure 9-5 Constructing Cisco VLANs. (*Source: Cisco*)

VlanDirector fully captures the physical topology of Cisco switches and routers within the campus LAN, as shown in Fig. 9.6, using an SNMP-based, reliable physical layer discovery process. VlanDirector uses these discovery services to provide logical VLAN views on top of the physical network topology, and to ensure that there is a reliable path between interconnected switches that have users assigned to a VLAN. This topology representation offers a highly reliable mechanism for configuring, modifying, and monitoring VLANs across multiple floors and buildings.

To set up a VLAN, the network manager fetches the form and specifies the type of LAN (Ethernet, token ring, FDDI). A protocol then gets the next set of available numbers for the map. The VLAN management domain allows the reuse of names and tagging in different locations as

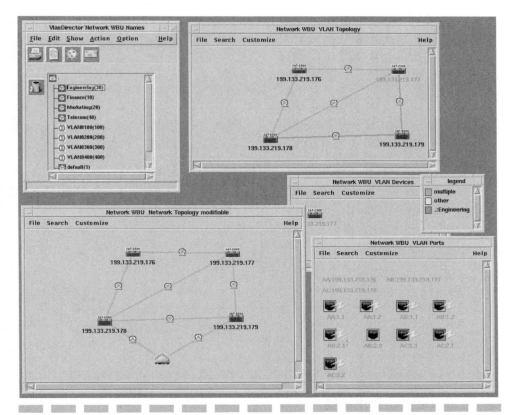

Figure 9-6 Cisco's VlanDirector: topology view. (*Source: Cisco*)

long as a router is in between. The naming tree advises of any duplication. VTP keeps the mapping tables unique.

VLAN Membership.   Membership can be

- Port-based
- MAC-address-based
- Network-layer-address based
- Policy-based (coming soon)

All of these can be established campuswide. To begin with, all users are put into VLAN 1. To add a VLAN, the screen shows a logical map overlaid on the physical map with click-on nodes revealing memberships, activity, and so on. On the left is a naming window with a form for addition of such details as name, LAN type, and purpose of VLANs. Clicking on the finished form also starts a VTP session, which then adds the new VLAN information to every switch. A table of all existing VLANs is also available at this screen. Adding members to VLANs is by drag-and-drop operation. (The ports then show up in blue.) The process can be repeated for the switches. For clarity, if the switches are not on the VLAN, an icon and represented links can be removed by drag-and-drop operation.

Moving a member across the campus entails use of the VTP, which knows about every VLAN and informs the switches. At present with Cisco VLANs, it is still necessary to reconfigure manually, although Cisco is moving toward an automatic process. If the user still belongs to the same router domain, then there is no need to change the network address. This also allows efficient use of class IP addresses. Dynamic addressing via DHCP is on its way. While this is complementary to MAC addresses, it is orthogonal to IP address use. Cisco feels that DHCP is of more use than VLANs. Finally, clicking on SHOW on the menu will enable the user to examine either devices, ports, the network, or VLAN views.

Virtual memberships under DHCP will be via the *Virtual Membership Policy Server*—by this means, the table of MAC addresses serves for finding locations and it approaches the DHCP server to discover an address for a user.

The configuration server, when released, will allow automatic membership by MAC or IP address or by real name once that name is attached to a MAC address.

Integration with CiscoView.   VlanDirector is fully integrated with CiscoView, Cisco's device management application. CiscoView can be launched by double-clicking on any of the device icons in VlanDirector's topology window. CiscoView, which is part of the CiscoWorks application suite, offers powerful graphic utilities for displaying chassis, interface, and port status information of a network's Cisco switches and routers. VlanDirector employs many of these graphical views to display VLAN status information by switch and port using color and legend indicators. (See Fig. 9.7.)

Drag-and-Drop Configuration Options.   VlanDirector uses a simple drag-and-drop GUI for assigning users to a VLAN and for reconfiguring these users to other VLANs. Ports are highlighted and moved across the screen with simple select functions and are dropped into the appropriate windows by releasing the select button on the mouse. Network managers can add new users to a VLAN, move users from one

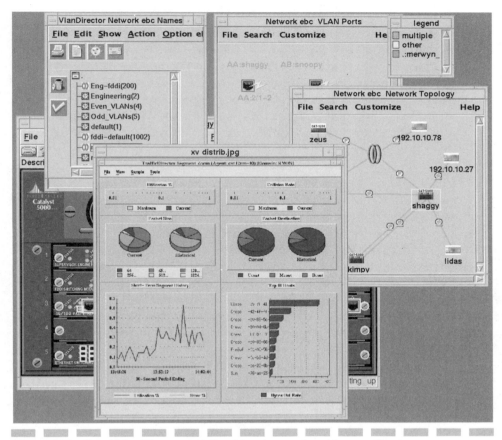

Figure 9-7   Cisco's VlanDirector and CiscoView integration. (*Source: Cisco*)

VLAN to another, or delete users from a VLAN with these drag-and-drop screen operations. To delete, drag the user to the trash-basket icon in the main system window. Multiple users can be highlighted as a group and added to a VLAN at the same time. VlanDirector provides several additional highlighting functions for adding and deleting users.

Discovery and Topology Mapping.   VlanDirector provides a campus view of interconnected Cisco switches and routers that is used for learning their connection relationship and for displaying logical VLAN topologies on top of the underlying physical network. A low-bandwidth SNMP discovery process is employed by VlanDirector to build the topology map. This discovery process works in conjunction with the

Cisco Discovery Protocol (CDP), which pulls in adjacent information about each campus switch and router. The topology map displays detailed information about the types of switches and routers, the types and number of interconnected interfaces, and the current configuration of the links, including the number of VLANs enabled. The discovery and topology mapping function is an integral part of the discovery process and is continuously updated by VlanDirector. When a configuration mismatch is discovered, VlanDirector automatically displays a window that flags the error and indicates the source of the problem.

Link Configuration Options. As part of its mapping and VLAN creation functions, VlanDirector automatically configures links between multiple switches when connecting ports that are geographically dispersed across a series of switches. Network managers can use VlanDirector in an automated link configuration mode that allows the application to perform link selections dynamically, or they can manually select links in a guided configuration mode. In its automated mode, VlanDirector automatically configures the appropriate interfaces as selected within the preference settings. These settings can be based on shortest-path selections, redundancy requirements, and bandwidth preferences or a combination of these options. Using VlanDirector in manual mode, network managers can fine-tune the VLAN configurations for applications such as disabling a VLAN across a link for security or performance reasons. As part of VlanDirector's drag-and-drop functionality, users can engage or disable links by dragging them to the appropriate window.

Report Options. VlanDirector provides users with detailed reports of configured VLANs both on screen and as printed reports. Network managers can obtain a listing of switches, ports, and links within highlighted VLANs. They can also request reports by switch and interswitch connections relative to VLANs configured, and can use VlanDirector's color-coded legends to display multiple VLANs concurrently. These reporting functions provide quick monitoring tools for understanding VLAN configurations that everyone in the network management organization can access.

Digital Equipment

Virtual LANs (VLANs) are an important part of Digital's enVISN architecture. Digital supports the new 802.1Q VLAN standard, but also offers a proprietary scheme at the moment and, when the standard is in place, will continue to offer additional features that the standard does not support. For example, where 802.1Q in its first pass supports only single spanning tree, Digital will support multiple spanning trees, allowing one per VLAN, to give better performance. Digital's main VLAN engine is the VNswitch, and this offers the standard VLAN. Digital will also offer protocol-based VLANs as an option in addition to the standard and later, MAC-based and policy-based VLANs, as the standard develops.

Digital intends to provide comprehensive VLANs by supporting a broad range of membership policies and addressing schemes within its enVISN architecture. Digital classifies its VLANs as

Class 1 VLANs—group a set of ports into a single broadcast domain (e.g., port switching). Class 1 VLANs can be used to dedicate certain ports to provide secure access to dial-out facilities.

Class 2 VLANs—group a set of end stations logically into a single broadcast domain across multiple hubs based on MAC layer addresses.

Class 3 VLANs—group a set of stations logically into a single broadcast domain based on a common network layer (i.e., subnet) address. This is also referred to as a *virtual subnet*. Class 3 VLANs are useful for protocols, such as IP, that bind the network layer address to a device via manual configuration or via an address server. Digital is the only major vendor that has added support for protocols other than IP from the start.

Digital allows VLANs to be formed from any combination of port, MAC, or virtual subnet members. By providing network managers with all three classes of membership policies and allowing these classes to be mixed together, Digital's enVISN architecture gives network managers maximum flexibility in building VLANs. For example, a combined class 1—class 2 VLAN allows the grouping of physical ports and MAC addresses, independent of protocol. This type of VLAN is effective for nonroutable protocols such as LAT and AppleTalk.

Digital's enVISN architecture supports VLANs across multiple current and future LAN technologies (Ethernet, token ring, FDDI, and ATM) and can create VLANs within a single switch in a stackable configuration, within a multislot chassis or across multiple hubs. Digital is one of the vendors which can combine ELANs and VLANs, trunking between them with up to 16 logical connections, although the first release is port-based only.

Digital's enVISN architecture addresses these consequences of scaling limitations by supporting enterprisewide virtual networks. A *virtual net-*

work (VNET) is a collection of VLANs interconnected with high-performance optimum routing. Each VLAN is defined using port, MAC address, or subnet-based membership policies. Communication between VLANs within a VNET is performed by high-performance multilayer switches (with integral routing) or by existing routers. A network with more than one VLAN requires routing between the VLANs; and, as the number of VLANs increases, so does the amount of routing required to enable inter-VLAN traffic. This routing introduces additional latencies and can become costly and difficult to manage. Digital's solution is to deliver high-performance optimum routing between VLANs by

■ Supporting one-hop switching service policy for very high-performance VLANs.

■ Combining centralized configuration and administration with distributed intelligence to build end-to-end virtual networks.

Employees who work on multiple projects and high-performance servers that are shared among multiple workgroups must belong to multiple VLANs, which can create problems. enVISN provides flexible service policies, one of which allows a station to be a member of more than one VLAN without breaking or otherwise compromising the VLANs involved. Very few vendors offer this facility—it is one of the options that add to the 802.1Q standard. Where it is offered, it is usually available only on MAC address LANs where complications of IP addresses apparently being in multiple places do not apply. Digital appears to have ironed out this problem.

Multilayer switches.   enVISN implements virtual networks by integrating VLAN and routing technology within multilayer switches. enVISN multilayer switches use ASIC chip technology to deliver cost-effective high performance. enVISN multilayer switches split routing into two distinct functions: forwarding and route computation. The *forwarding function* is performed within the switch fabric, which uses custom hardware and fast memory to cache forwarding entries. The *routing function* is implemented in a separate subsystem, using off-the-shelf processors and low-cost memory. Because the forwarding function is separated from the route computation function, which can work in the background, enVISN multilayer switches provide very low-latency frame and cell forwarding, alleviating the latency bottlenecks created by traditional backbone routers.

enVISN allows inter-VLAN connectivity to be performed with existing routers where each router port belongs to a single VLAN. While this solution reuses existing routers, it will not provide optimum inter-VLAN routing. A packet may traverse to a router-connected VLAN switch even though both the source and the destination are in different VLANs but on the same switch.

Distributed Routing. Digital's enVISN architecture employs standard distributed routing algorithms in multilayer switches to achieve a low-cost solution with low latencies, scalability, and robustness. Other architectures being promoted today align with Digital's view that the traditional routing function should be split into two distinct functions—route determination and frame forwarding—and that the forwarding function should be distributed to and become integral with the switching function. However, Digital's enVISN architecture goes one step further by distributing the route computation function to the switches as well. Digital believes that centralized route computation schemes can be costly to deploy, complex to manage, and costly to use for increased traffic load support. A centralized routing scheme may also need additional route servers for scalability. Proliferation of route servers in a centralized routing architecture also presents data synchronization problems among the servers. enVISN therefore uses a distributed routing system to provide a fault-tolerant environment in which performance and cost scale gradually as the network grows.

Another fervent Digital credo is that latency is becoming more important in today's client/server computing environments, especially latency incurred during a transaction. Distributed routing delivers lower latencies than do centralized route server schemes—especially during initial connection setup. Central routing schemes cannot match the low end-to-end latency of enVISN's distributed routing system, says Digital. With enVISN, it hopes to combine flexibility in creating workgroups with a high degree of control in defining the environment for those workgroups by offering a wide range of service policies. These service policies allow network managers to fine-tune their networks and allocate resources fairly to users in the following ways:

Access to specific network services. Once a user is authenticated, Digital's enVISN will assign privileges to that user. These privileges create a service policy domain for that user, who can then communicate with any member belonging to the same service policy domain (workgroup).

Bandwidth utilization. A service policy domain can specify the maximum peak bandwidth utilization for a workgroup.

enVISN service policies are, unusually, flexible enough to allow for overlapping VLANs. This makes it possible for a station to be a member of multiple workgroups concurrently. For example, multiple workgroups could share a common, high-performance server. This server would be a member of multiple VLANs. Service policies can be applied across multiple VLANs. The default policy for interconnecting the three VLANs may not support the specific requirements needed by this cross-functional bid team. A policy membership for the bid team could be implemented to allow members to set up a videoconference and share a working *Request for Proposal* (RFP) document. The service policy for this domain would include low-latency quality of service to support videoconferencing and controlled access to the RFP document. enVISN service policies are administered centrally and distributed only to those devices that need to support the bid team, thus simplifying policy administration.

One-Hop Switching.   enVISN enables the formation of very high-performance workgroup VLANs through a unique service policy that Digital terms "one-hop switching." This uses the routing and virtual LAN intelligence embedded in multilayer switches to make switching decisions locally rather than relying on external devices such as central routers or route servers. The result is reduced connection setup time and faster data transfer, for improved network performance and lower latency. In a typical ATM network, the logical router topology is constructed independently of the physical ATM switch topology. It is not atypical for a path between two points to include multiple routers. In ATM environments, each time a packet traverses a router, it would cross the ATM "cloud" boundary (i.e., "hop"), be segmented into cells, and then be reassembled. One-hop switching eliminates the multiple "hops." One-hop switching utilizes VLAN and ATM switching topology knowledge contained in each enVISN multilayer switch to create a direct-switched connection to the destination. Thus, a packet takes only a one-hop path between its source and destination. Even though the one-hop path may actually traverse multiple ATM links and switches, packets are segmented into cells and then reassembled into packets only once over this path, minimizing end-to-end latency.

Network Management.   Digital's enVISN network management centralizes policy administration while distributing enforcement. The advantages of this approach are

- *Simplicity.* It is easier to administer networkwide policies than to configure individual devices.

- *Consistency.* Policies are checked before they are distributed to devices for enforcement.

- *Resiliency.* Enforcement is distributed to each device, leaving no single point of failure.

- *Economy.* Only a small subset of the total policy database is needed on each device.

- *Efficiency.* Policy distribution is done as a background task, minimizing potential bottlenecks.

The network manager uses two applications to manage the virtual network: (1) the element manager, for monitoring, configuring, and troubleshooting individual devices; and (2) the VNET manager, for policy management and monitoring of the VNET. The VNET manager provides the interfaces to policy management, topology mapping, events monitoring, and performance statistics.

Policy management includes

- *Membership*—who is in a given service-defined VLAN or workgroup

- *Security*—a combination of membership and service that provides access control on a per client basis

- *Services*—a set of constraints, such as a specific amount of bandwidth, that is applied to a given VLAN and workgroup membership

The VNET manager also provides a graphical interface to the following three VNET monitoring functions:

- *Topology mapping*—a logical and physical mapping of the devices on the VNET to enable the network manager to manage via the logical topology (i.e., VLAN membership) but troubleshoot via the physical topology (e.g., FDDI ring map)

- *Event monitoring*—collection and analysis of events generated by the intelligent devices and the generation of an appropriate set of action routines

- *Performance monitoring*—analysis and display of statistics that are generated by intelligent devices

The VNET manager converts policies into topology rules, which are then passed to the VNET configuration server. The VNET configuration server automatically calculates the VNET topology and performs consistency checks on the VNET policies. Once verified, the configuration server distributes configurations to the individual devices in the VNET for implementation.

The VNET monitoring functions get data from autodiscovery and RMON services or the VNET database. Autodiscovery and RMON services run in the background, collecting data from intelligent agents and passing it to events and performance monitoring and/or storing it in the VNET database.

clearVISN VLAN Manager v1.0. Digital's current offerings also include clearVISN VLAN Manager (see Fig. 9.8)—a graphical, SNMP-based application that enables network managers to configure and manage port-based VLANs as implemented within the enVISN architecture on the DECswitch 900EF (FDDI) and PEswitch 900TX (Fast Ethernet) product. VLAN Manager complements clearVISN MultiChassis Manager by using a logical approach to effect moves, adds, and changes across the organization.

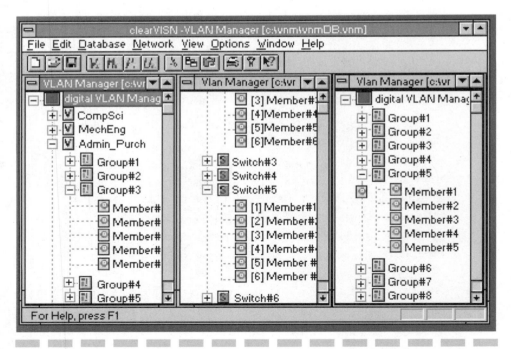

Figure 9-8 Digital Equipment's clearVISN VLAN Manager. (*Source: Digital Equipment*)

With the clearVISN VLAN Manager, VLANs can be created by simply defining collections of member ports selected from a directory of switches and switch ports provided in a VLAN manager window. The VLAN manager automatically sends the appropriate SNMP information to the list of switches affected and assigns the appropriate ports to that VLAN. Members can be attached to VLANs by simply dragging a switch port to a defined VLAN name. Groups of members can then be moved from one VLAN to another by dragging the group name or any member of the group and dropping it on the desired VLAN name. A key application in Digital's clearVISN network management solution, the VLAN Manager can be integrated with other clearVISN applications through the clearVISN Core.

FORE Systems

ForeView v4.0 network management software manages ForeRunner ATM networks through a variety of network management environments. The devices ForeView can manage include ForeRunner ATM switches and ATM network adapters, PowerHub LAN access products, and later this year, CellPath WAN access multiplexers. It is fully integrated with the HP OpenView, and Solstice SunNet Manager platforms or runs standalone. ForeView is currently available for all of the following operating systems: Windows NT v3.5, Sun Solaris 2.3 or 2.4, SunOS 4.1.3, SGI 5.3, and HP-UX 9.x. Integration with the industry-leading network management platforms reduces costs and simplifies operator training.

ForeView network management software has a wide range of features that help network managers implement and maintain their ATM network. The 10 most commonly used features are

- ATM Forum UNI 3.0 automatic discovery and mapping features
- Configuration and monitoring functions
- Virtual path/channel tool
- Channel tracer
- ForeView inventory application
- ForeView graphing tool
- Scripting utility
- Logging utility
- Full on-line help
- Virtual LAN manager

Here, we will take for granted the extensive features of network management that FORE Systems provides, to concentrate on what FORE calls the *virtual LAN manager*, although in fact it should more precisely be called *ELAN manager*, since it deals with Emulated LANs under LANE. The VLAN manager is available as an optional extra with ForeView.

The VLAN Manager Tool. As part of FORE's release of the ATM Forum-compliant version of LAN Emulation (LANE v1.0), ForeView v4.0 includes a virtual LAN application that simplifies the configuration of ELANs by enabling users to perform drag-and-drop operations to graphically specify ELANs and the membership of hosts within those ELANs. Specifically, the ForeView VLAN manager provides the following functions:

- Retrieves and deposits LECS configuration file on LECS
- Provides screens to configure parameters such as MTU size for each VLAN
- Provides commands to add, move, and delete (drag-and-drop) nodes between VLANs
- Provides commands to create or delete VLANs
- Prompts for LES NSAP address for newly created VLANs

Figure 9.9 illustrates the VLAN manager tool.

Newbridge Networks

Following is a description of the former UB Networks VLAN—it is expected that Newbridge will take this up as it is well advanced and Newbridge has nothing equivalent—although, of course, as Newbridge is an ATM vendor, it supports ELANs. Look out for better integration with LANE to come as an early improvement, and a speeding-up of the more advanced IP VLAN version, to match up the old UB products with Newbridge's VIVID range.

UB Networks had an interesting history when it came to VLANs. It was the first company to attempt to implement the idea of a virtual network. It invented its own scheme—Virtual Network Architecture (VNA)—which was available only on a single chassis, the DragonSwitch (no longer available under that name, although still around as part of other switches, notably the GeoLAN/100 and GeoLAN/500). The DragonSwitch had 80 ports, and within that limit, it had the possibility of

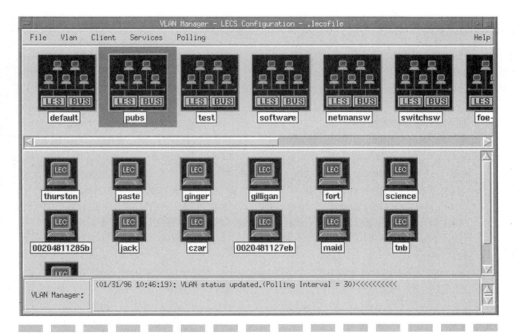

Figure 9-9 FORE Systems VLAN manager tool. (*Source: Fore Systems*)

overlapping VLANs. Because of the clear and simple colors and schema, it was easy to move users around. In 1994, VNA2 was developed, using the same interface, based on the 802.10 standard. This was originally available on FDDI; now it is also offered for ATM and Ethernet. As yet, it is not ready for Fast Ethernet. The reason for FDDI being first off the block was that UB felt it was essential to get the backbone in place first, and FDDI was, at that time, the most popular technology for backbones in LANs. With this VLAN, it is not possible to have a full heterogeneous environment—there can be no token rings mixed in. UB's PlusBus (from the original DragonSwitch) supports only a two-port-per-card bridge for token ring, which is not of a sufficient order of magnitude. While switching is essential for a true VLAN, this VNA2 can be used for microsegmentation on a shared network. In the GeoLAN/500 chassis, there are two buses: PlusBus and GeoBus. PlusBus can carry only 320 Mbits/s, while the real VLANs sit on GeoBus, which is a full ATM base with 10-Mbit/s throughput.

The most advanced developments are in the GeoLAN/500 (although it is now also possible to trunk between that and the older GeoLAN/100). The Port Mobile Ethernet Concentrator (PMEC) has twenty-four 10-

Mbit/s ports, each of which can be assigned to any segment. Each segment is bridged by a central switch, the CSSw 12-port switch (this is essentially a DragonSwitch). It is possible to have up to 264 PMECs in a chassis, giving a total bandwidth of 120 Mbits/s. PMEC gives slow, flexible, port switching.

GeoBus Subnet.   This is a new addition to the existing buses on the GeoLAN switches. It uses high-speed, frame-switching, Fast Ethernet, 16-port, self-configuring $^{10}/_{100}$ cards and/or 24-port 10Base-T switching cards—these can interoperate. This forms one large flat network unless configured otherwise, making a VLAN environment essential. There is also a Fast Ethernet concentrator and a two-port ATM or FDDI (and there will be Gigabit Ethernet using the 800 Mbits/s fiberchannel later in 1997) uplinks. The GeoBus backplane speed is 7.2 Gbits/s with speed between any two slots of 1.2 Gbits/s. It has four-group RMON on any port with software-based port mirroring. On that mirror, it can hook up to UB's EMPower management package, giving the full nine groups. It has the PMEC plus the PacketDirector ASIC for Ethernet and Fast Ethernet. With this architecture it enables any of port, MAC address, or IP subnet VLAN.

MultiLAN.   A so-called MultiLAN is then formed with up to 264 workgroups, which can be overlapping. The PMEC system is here today and working, and is inexpensive. In UB's words, it is "not rocket science." It is limited to only 12 physical segments. If the DragonSwitch is used to its full potential, it is possible to get 65,000, but this is more expensive and each one is limited to just 80 ports, rather than the 264 of the PMEC.

UB, now within Newbridge as part of the Entr@Networking Division, is in the process of setting up new facilities for its VLANs. The first development is the Virtual Network Visualizer (VNV), which, when it arrives, will make management easier and more automated. The first elements will be in use from early 1997. The Venn diagrams that show overlaps (as in Fig. 9.10) and Pipe diagram (also in Fig. 9.10) that maps users will be among the first elements available.

UB's main new thrust, however, is in putting together the VLAN built to 802.1Q. This became necessary with the need to get across layer 3 breaks in the network that IP-based 802.1Q networks could do.

Configuration of MultiLAN is as follows. To set up a VLAN, the administrator has two lists—names and (color-coded) workgroups—and clicks on the intersections to assign names to the groups. Then MultiLAN automatically transfers to the graphical Venn diagrams. Within this interface, it is also possible to assign other types of groupings.

Figure 9-10
Newbridge Networks
Virtual Network
Vizualizer. (*Source:*
Newbridge
Networks)

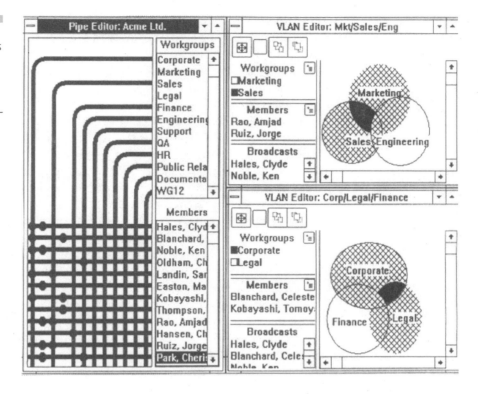

Traffic Tuner. This is a simple but helpful way of breaking up a network into virtual units by enabling the movement of people on the basis of performance needs. For example, if five segments on the backplane are used too heavily, Traffic Tuner can reassign on that basis. This does not fit directly into UB's VNA scheme but can be used alongside it. It works over the PMEC, giving port-by-port or segment-by-segment traffic display. On the PMEC, with its limit of 12 ports, it is not too difficult administratively to set up, say, two ports to each workgroup and adjust later.

The New VNA. The world is moving toward subnet mapping for scalability, and the limitations of Traffic Tuner and MultiLAN give them a brief future. UB's new VNA is on its way with the new platform, the GeoBus (see section on GeoBus subnet above).

 This is a port- and MAC-based scheme with the benefit of IP-subnet-based VLAN definition. (In this context, a *subnet* means a VLAN for move, add, and change purposes.) On setting up or moving a user, no management is needed. The subnet address and VLAN are tightly coupled, taking the load off the network manager in moving people around manually. On the other hand, manual intervention, if done incorrectly, can cause major problems, so in this case it is advisable to allow the subnets to be the same as the VLANs. Of course, this has ramifications: (1) it is impossible to have overlapping VLANs, and (2) one cannot have workgroups across routed networks. Both of these features were possible in the earlier scheme. (To belong to two VLANs, a user will have to go through a router, which is slow and expensive, and which introduces an artificial break in the network.) However, it is easier to set up, and it scales for large network use. At present Newbridge is not at all sure how easy it will be to make later changes.

UB Networks allowed for the use of routers in this landscape, or switches with routing abilities on them. As Newbridge, it is introducing an embedded scalable routed network architecture. It is simple, with only RIP and IP protocols supported. *Address Resolution Logic* (ARL), implemented in the PacketDirector ASIC, looks at the first 128 bits of the packet in a data stream to see if it is IP and a known (i.e., cached) address; if it is, it goes on the "fast path." If it is not, it uses RIP and the "slow path." (This is analogous to both IP switching and ATM's MPOA.) If both the origin and the destination are in the same VLAN and the VLAN is IP-based, then the traffic is switched straight through; if it is not IP-based, it is switched to an external router for DECnet, IPX, or other protocol. Newbridge is somewhat reluctant to use the term *layer 3 switching* because different companies use it in different ways. For example, Cisco, which has a router-server for address resolution, calls it *adaptive subnet switching*, but it is essentially switching using layer 3 information based on automatic IP address assignment, no matter what is plugged into the network. In summary, then, intrasubnet traffic is switched; intersubnet traffic is routed. The next step will be for UB to enact this in hardware-based IP routing, rather than the mix of hardware and software as at present. With virtual IP, the logical and physical networks will be independent. A DHCP server will perform the assignments and addressing, and then the user will belong to a group. The server increments IP addresses one by one. However, although this is fine until the VLAN boundary is crossed, no one is yet sure how this will work in practice—how additions and deletions will work, for example, or how the boundaries of different-size VLANs will be arranged.

VLANs with Hubs.   Although VLANs are used essentially for switched networks, it is perfectly possible to employ stackable hubs in the workgroups (although the granularity will be reduced), and they should work in the same way. In the VNA, stackables will implement a subset of the possible facilities and manual intervention will be necessary for all except the new GeoLan500. Newbridge claims that there, everything, on all platforms, will be homogeneous, automatic, graphical, and managed.

Management.   UB Networks had an overall network management system called NetDirector, based on the Hewlett-Packard OpenView management platform. VNV and Traffic Tuner are implemented in NetDirector. There is also NetDirector@Web—which UB claims was the first Web-based manager. It uses Netscape or any other standard HTML Web client, or a Web browser. Here EMPower becomes a Web server running Java, the language based on a cut-down C++ used for Internet applications. EMPower is a management module from UB for monitoring. With it, it is possible to get a physical view, see traffic patterns, and more. Newbridge is taking up all these management platforms and making them the basis for management across all the Entr@Networking products, both Geo-based and VIVID-based. ACC routers are also sold by this group, and NetDirector is expected to be expanded to include these, too.

Xylan

AutoTracker, Xylan's second-generation virtual LAN implementation, introduces automatic policy-based virtual LANs to any LAN, MAN, or WAN network, and is one of the few VLANs to be policy-based. When implementing a switched network with AutoTracker, the first step is to create AutoTracker groups—a collection of physical ports on one or more OmniSwitches. One spanning tree operates for each AutoTracker group. Xylan has designed AutoTracker to be simple to set up.

It's reasonable to think of AutoTracker groups as functionally separate switching systems contained within the same set of switches. An individual AutoTracker group can have any combination of AutoTracker VLANs within it. Each group is identified with a 16-bit networkwide identifier. An OmniSwitch network can have 65,536 groups, each of which can be port-based virtual LANs or other AutoTracker VLAN types. Any OmniSwitch can support as many as 96 active port-based VLANs or groups at one time.

Each AutoTracker group has its own set of policies. The policies contain a set of virtual LAN criteria which operates only within that Auto-Tracker group. No traffic can ever move between AutoTracker Groups except through routing.

AutoTracker is capable of supporting the following VLAN types:

- Port-based
- MAC-address-based
- Layer 3 (IP, IPX, AppleTalk or DECnet)—based
- Protocol-policy-based (protocol-based)
- Multicast-based
- Policy-based
- Authenticated-user-based

Authenticated user VLANs is something that only Xylan offers, and has not been discussed before. This type of VLAN is purely a higher level of VLAN security on top of any other Xylan VLAN type. It takes network security to this higher level by requiring users to be authenticated by a server to gain access to network resources—in other words, it adds a firewall. For instance, an authenticated user VLAN can be made up of the users from the finance department. To access the resources in the finance department, the user must pass through an authentication firewall built into the OmniSwitch and PizzaSwitch. (See Xylan's Authentication Server in Fig. 9.11.)

An AutoTracker virtual LAN can include, and connect, devices with a variety of MAC types. In a single virtual LAN, you can include Ethernet and token-ring workstations connected to FDDI servers. Any AutoTracker virtual LAN can span an entire building or campus. Likewise, members of the same virtual LAN can be connected to any number of PizzaSwitches across a Fast Ethernet, FDDI, TP-PMD, or ATM.

Features of Auto Tracker

1. An AutoTracker Virtual LAN can include any combination of media types. For example, a single virtual LAN can include end users at one location connected by Ethernet, a server running FDDI, and users running token ring at a remote site connected by frame relay. AutoTracker even lets network managers have ATM included in a virtual LAN.

2. Any given virtual LAN can include any number of ATM Emulated LANs (ELANs).

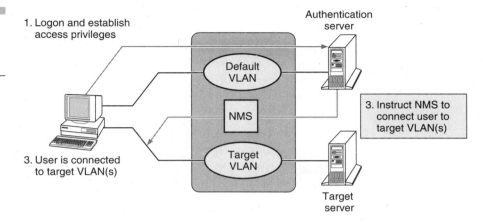

Figure 9-11
Xylan authentication
server. (*Source:*
Xylan)

1. Logon and establish access privileges

Authentication server

Default VLAN

NMS

3. Instruct NMS to connect user to target VLAN(s)

Target VLAN

3. User is connected to target VLAN(s)

Target server

3. Any given virtual LAN can span an entire building, campus, MAN, or WAN; members of a virtual LAN can be connected to any number of OmniSwitches and/or PizzaSwitches.

4. AutoTracker virtual LANs can be established automatically according to a set of policies configured in the network management system (OmniVision). AutoTracker policies include MAC address, layer 3—based policies, custom policy, protocol-based policies, and multicast VLAN-based policies.

5. A single-switch port can support multiple virtual LANs. This feature allows a hub to be connected to a switch port with up to 32 VLANs per port. The PizzaSwitch and OmniSwitch automatically sort the devices attached to their ports into virtual LANs according to the policies that have been defined.

6. A single device can belong to more than one virtual LAN (up to 32). This feature provides optimal support for servers which can have users on a number of different virtual LANs.

7. A subnet can span multiple switches; a network manager doesn't have to create a separate subnet for every switch or switch port. The result is more efficient use of subnet addresses.

8. Devices are automatically tracked within an AutoTracker group as they move around a network. Policies are applied consistently within an AutoTracker group, so a workstation or server automatically stays in the same AutoTracker virtual LAN as long as it's moved within the same AutoTracker group.

9. AutoTracker groups can be established across an FDDI backbone using Xylan's trunking protocol. Multiple FDDI rings can be used to connect an OmniSwitch network together. The OmniSwitch can also send information directly to FDDI-attached devices in native form on the same physical-MAC interface which is carrying trunking streams between OmniSwitches.

10. AutoTracker groups can be established across an ATM backbone using Xylan's trunking protocol, thereby allowing OmniSwitches and PizzaSwitches to be interconnected across any ATM switch or switch fabric using either permanent virtual circuits (PVCs) or switched virtual circuits (SVCs).

11. AutoTracker groups can be established across an ATM backbone using ATM LANE or RFC1483. In fact, one portion of a virtual LAN can use Xylan's trunking protocol, while another portion of the same virtual LAN can use ATM LANE. This permits flexibility in combining equipment and allows the easy migration of a network as hardware is upgraded.

12. AutoTracker groups can be established simultaneously across an ATM backbone and an FDDI backbone. This is of value when a company has one or more FDDI backbones in place and wants to migrate to ATM in the future. The FDDI network can be left in place to carry a portion of the load and serve as a "hot backup"; meanwhile administrators do not have to change AutoTracker groups while the migration takes place.

13. Separate spanning trees are maintained for each AutoTracker group. This provides a high level of flexibility when creating VLANs.

14. AutoTracker can simplify management of source-route networks by allowing virtual rings to be created. Multiple physical token rings can transparently connect to one logical token ring. This reduces the hop count. In fact, a virtual LAN adds only one hop in a network. Virtual rings can extend across both an ATM backbone and an FDDI backbone. Standard source routing is performed between virtual rings, providing complete interoperability with existing bridges.

Broadcast and Flooding.   When most switches receive a frame with a MAC destination address that it does not recognize (including broadcast frames), they simply flood that frame to all ports. An OmniSwitch or PizzaSwitch network uses more sophisticated techniques to handle broadcast frames and frames with unknown destination addresses. Such a frame is forwarded to a number of ports in the same AutoTracker group from which it originated—but not necessarily all the ports. There are three possibilities:

1. If the OmniSwitch has identified that the originating device belongs to one or more virtual LANs, then the frame is flooded to all ports that are members of those virtual LANs.

2. If the OmniSwitch has not yet identified that the originating device belongs to a virtual LAN, but it has identified that other devices on that port belong to one or more virtual LANs, it will flood the frame to all ports which have members of those virtual LANs.

3. If the OmniSwitch has not yet identified that any device on the originating port belongs to a virtual LAN, the frame is forwarded to all ports within the group. At the same time the *management processor module* (MPM) analyzes the frame, and if possible, adds virtual LAN assignments for that port.

In no case will a frame with an unknown MAC address destination be forwarded to any ports which are outside the group in which the frame originated. This keeps each AutoTracker group completely isolated.

Device Movement. When a workstation or other device moves from one location to another, it will attach to the network on a different switch port. If it moves far enough, it will even be attached to a different switch. The OmniSwitch network automatically learns the virtual LAN assignment(s) of the device and keeps it operating just as it had before the move.

AutoTracker enables a network device to be moved to a different physical location within an AutoTracker group, yet remain connected to the Virtual LAN defined for the device's logical workgroup. To use Auto-Tracker, an administrator must first decide whether AutoTracker groups are needed and if they are, what those groups will be. Next a network manager needs to determine the policies which will be applied to the group(s). The policies that a manager defines are then downloaded to individual switches using a network management station. AutoTracker then places users into virtual LANs according to those policies. Subsequently, any movement within AutoTracker groups are tracked.

Data Transfer. All the policies that an administrator implements (port-, MAC-layer 3—protocol-, multicast-, or custom-defined) that apply to an OmniSwitch or PizzaSwitch are stored in the MPM. Unicast frames from unknown devices, broadcast frames, and multicast frames are all sent to the MPM. The software examines the entire frame to determine membership in a virtual LAN for that source MAC address. It is important to note that during the switching process, a MAC frame will be switched according to the originating station VLAN membership. Broadcast or multicast frames will be processed to determine VLAN membership; unicast frames are examined only the first time they are seen.

Dynamic Changes within a Group.   When a device that is a member of an AutoTracker VLAN moves to a port on a switch where the device's VLAN is not active, the new Xylan switch reconnects that device to its virtual LAN after receiving and analyzing the first frame from the device that moved. In contrast, if a virtual LAN has been active, and for some reason all stations that are part of that AutoTracker virtual LAN cease to be active, the virtual LAN is shut down on that switch until a device which needs to access it is connected.

Optimized Device Switching Ports.   *Optimized device switching* occurs on a port connected to a dedicated LAN segment with a single MAC device attached. No spanning-tree bridge protocol data units are sent along this type of port, and the port itself is always in a forwarding state. With this type of switching, no flooding of unknown unicast frames occurs. If more than one MAC source address is detected, or inbound spanning-tree bridge protocol data units are detected, this port type automatically changes to a bridge port and an SNMP trap is sent to a network manager. Optimized device switching ports are a Xylan innovation which allows the OmniSwitch and PizzaSwitch limit traffic on individual ports.

Trunk Ports.   Trunking is required to support AutoTracker group VLAN distribution across multiple OmniSwitches interconnected by LANs or ATM networks. Trunk ports are unique in that they encapsulate data with a header so that separation of the data from multiple virtual LANs can be maintained over a common interconnection. FDDI ports can be configured as a trunk port, and an ATM port can support multiple trunk ports. AutoTracker Groups/VLANs can be extended across a backbone using trunk ports. Trunk ports are configured by the user; trunking carries the original CRC and framing information.

The OmniSwitch can simultaneously run trunking within an AutoTracker group while forwarding frames to end devices on high-speed networks such as FDDI and ATM. Frames destined for network devices on an adjacent LAN are forwarded in the native encapsulation for that LAN. Frames forwarded over an FDDI VLAN trunk are SNAP encapsulated. In the case of an ATM network, separate virtual circuit connections are established for the virtual LAN trunk and the ATM end stations communicating with the OmniSwitch.

FDDI Virtual LAN Trunking Protocol.   The VLAN trunking protocol is SNAP-encapsulated and uses unicast addressing. The switch transmits trunked frames with a unique source MAC for each Auto-Tracker group or VLAN extended over FDDI. When the source MAC address and the SNAP field are examined, a frame can be identified as a trunked frame and its AutoTracker group membership can also be identified. When the frame is learned, the source MAC address of the remote OmniSwitch is maintained. When flooding traffic, a general-purpose unique unicast is used as a destination MAC address for forwarding of unknown unicast, broadcast, and multicast traffic. When a known unicast is transmitted, the MAC address of the remote OmniSwitch is used. This enables the protocol to operate together with traffic from other protocols over shared LANs.

ATM Virtual LAN Trunking.   The OmniSwitch or PizzaSwitch overlays the trunking functionality on virtual circuit connections (VCCs) established between the OmniSwitches and PizzaSwitches within an ATM network. Traffic can be separated by using unique VCCs for trunked traffic. The AutoTracker group trunking protocol does not utilize the multicast services of the ATM network such as the Broadcast and Unknown Server (BUS).

The AutoTracker group identifier and name are globally recognized by each OmniSwitch in the network. The AutoTracker group trunk protocol carries the group identifier in the encapsulated frame as it is transmitted over the trunk. All OmniSwitches and PizzaSwitches which contain the matching AutoTracker group will process the frame as it is received from the trunk.

OmniVision Network Management Suite.   Xylan's OmniVision application suite is an entire set of network management applications which allow easy configuration, monitoring, and diagnosis of a network. OmniVision automatically collects MAC addresses and allows network administrators to associate a MAC address with a physical device by name. The first two applications available for the OmniVision management suite are SWITCHmanager and AutoTracker.

The *SWITCHmanager* application provides complete configuration, monitoring, and diagnostic information for a PizzaSwitch-OmniSwitch network. It uses detailed graphics for real-time reporting of network performance, alarms, and configurations.

AutoTracker, a graphical application that allows the creation and management of policy-based virtual LANs for PizzaSwitch and OmniSwitch, enables the user to view the relationship between groups, virtual LANs, and devices across the entire network. Drag-and-drop features with setup wizards facilitate the setup and maintenance of VLANs.

OmniVision's application suites can be used as standalone applications to provide a management system for Xylan's product family. For powerful enterprise management of an entire network which may include a mix of different vendors' equipment, OmniVision applications can be integrated with a variety of enterprise network management platforms. These platforms include HP OpenView for Windows, HP OpenView for UNIX, and Sun's Solstice SunNet Manager (versions for IBM's NetView for AIX and Cabletron's Spectrum are under development).

Recommendations

Because of the lag time necessary for some vendors to incorporate the frame format specification and most organizations' desire to have a unified VLAN management platform, VLANs will, in practice, continue to retain characteristics of a single-vendor solution for some time, which has significant ramifications for deployment and procurement of VLANs. Department-level procurement for LAN equipment, particularly in the backbone, is not practical for organizations deploying VLANs. Purchasing decisions and standardization on a particular vendor's solution throughout the enterprise will become the norm, and price-based product competition will decrease. The structure of the industry itself may also shift in favor of the larger networking vendors that can furnish a complete solution across a wide range of components.

Looking at the variety of offerings on the market today, it is easy to get confused. It will remain this way until things settle down with the new standard ratified and all vendors complying with it. Also, as the market settles and it is more obvious which are the advantages and disadvantages of the different approaches, the products will become increasingly homogeneous.

The main rule of thumb in buying is that you don't go out to choose a VLAN—you go out to buy switches, and the VLAN that comes with them is the one you get. Nobody should worry about the VLAN yet—the products are changing and improving all the time. What you should do is bear in mind the switch vendor's attitude toward VLANs—has it got a strategy? A product? If it has, then, good—but is it workable? Flexible? If it does not have a product or even a strategy, then you are unlikely to be able to implement VLANs in your network with its switches for a long time to come. The VLAN schemes from the vendors outlined here vary—they may be good, bad, practical, or far-fetched, but they all show one thing: willingness.

VLANs are a very new idea and at the moment a lot of hype is attached to them, and everybody thinks they will solve all their problems. As you can see, with the products outlined here, there are tradeoffs, some things get fixed, but a lot of other problems are created instead. One example is network manager's time—VLANs are supposed to save it, and VLANs can (some of them) automate some of the network manager's tasks and take the strain off the network, which will help in that it won't break so often. So far, so good. On the other hand, the network manager will have to learn the new tricks needed to work it (and the new versions as they come out), spend a lot of time to "save time" in setting up the various VLANs, and explain to the users why they can see only certain data, and so on. It's not easy. Just keep in mind these recommendations when you go shopping:

- Buy all your switches from the same vendor if you want to run VLANs.
- Buy for the strength of the switches, not the VLAN features.
- Buy from a company that does have some ideas about VLANs now if you think you will ever need them.

When planning your VLAN, take great care with any mixture of layer 2 and layer 3 structure so as to avoid broadcast loops and other anomalies. It is a good idea at the planning stage to build in a lot of over capacity, both to avoid having to add to the new structure piecemeal in the future and to include the rapid growth that may ensue from having a much more efficient network, as for example, when all the remote workers suddenly want in.

When it comes to installation, it is necessary to ensure that all the LAN switches and management tools are up and running before deployment. That way it is possible to test and configure each device in isolation as it is installed, rather than having to debug the whole network. It is a good idea to run the management tools under a larger "umbrella" NMS (network management system) such as Hewlett-Packard's Open-View or Sun's Solstice SunNet Manager to give a single point of control to simplify integration and one dial-in number for off-hour coverage. Cabletron's Spectrum is also a good choice, especially if buying Cabletron VLAN and switches—it is increasingly viewed in this category rather than as a simple element manager. It is also a good idea, if using an element manager such as 3Com's Transcend or Bay Network's Optivity, to make sure that the VLAN you have will work alongside. If your switches and VLANs come from the same vendor as the element manager, then you have no problems—it might not be integrated at the moment, but it will be before long.

There are some features that any good VLAN will have

Multiple media types. An example of this could be 10-Mbit/s Ethernet workstations that may access a server connected to a switch by 100-Mbit/s Ethernet, Gigabit Ethernet, or ATM. The VLAN should be able to handle any media type currently used as well as any media that may be added.

Switches and hubs. Although we are moving toward a fully switched world, it would be useful if the older hubs could be part of the VLAN. They can then be connected to a port on a switch supporting multiple virtual LANs. This way, even though the end stations attached to the hub don't have a dedicated switch port, they can still connect to VLANs.

Layer 3 support. To move beyond the single location, a router is used at present. A future-proofed VLAN will be able to carry VLANs across the currently routed links.

Servers supporting multiple VLANs. With advanced VLAN implementations, servers are part of multiple VLANs. Information doesn't have to pass through routers or be broadcast across the backbone. Broadcast information is contained within each VLAN and doesn't spread needlessly throughout the network.

Multiple-switch networks. Virtual LANs are of little benefit if they can't span multiple switches, and as yet some can't. The need for switching is greatest in midsize and larger organizations, which usually need multiple switches.

FDDI backbones. FDDI is a proven and reliable high-speed backbone technology that many organizations have used successfully for years and do not want to displace. Not all vendors' virtual LAN implementation support FDDI. If integrators want to use their existing FDDI media, they need to be sure that the VLANs a switch offers has support for FDDI as well as emerging technologies.

ATM backbones.   ATM is quickly becoming a strong alternative to FDDI as a backbone technology, and may become widely used for server connectivity. Switching options which may be suitable now may not be optimal in the future. Generally the more powerful and flexible the standard, the farther it is from final standardization and widespread implementation. It's important that a virtual LAN mechanism be able to operate efficiently across ATM backbones.

Automatic adds, moves, and changes.   In many networks it's common for devices to move frequently within a building or campus. Network administrators should be able to assign devices to virtual LANs and never have to reassign them, no matter where they move to.

High-speed operation.   VLANs should not affect a network's speed, although at present they frequently add latency due to relatively inefficient lookups.

Which Vendor to Go For?

If, having used all the general recommendations discussed above, you are still undecided as to which vendor's VLAN to buy, it is time to look at the several surveys that have appeared evaluating the current crop of offerings.

In October 1996, *Datacommunications International* magazine published surveys and in May 1997, tested the VLANs from most of the major vendors. In the survey it found that

- All vendors offered the basic facility of port-grouping and protocol-based VLANs, and that all except IBM also had MAC address grouping.

- All except Bay Networks had protocol addressing, all by IP, but Agile, Cabletron, FORE, IBM (due late in 1997), and Xylan also by IPX. AppleTalk was supported by Agile, Bay, Cabletron, and FORE.

- The number of concurrent VLANs supported varied from 32 to 64,000, but at this stage this is not something to worry about—while it may be a good idea to run limited VLANs now, by the time it becomes critical to run them across the enterprise, all will be scalable.

- The addition of RMON groups was very patchy—from none (IBM) to all nine [Case Technology, now Intel, plus Cabletron and UB (now Newbridge) by means of the roving probe]. Again, this is something that will rapidly improve until at least the four main RMON groups of Events, Alarms, History, and Statistics are covered.

Cisco and Digital Equipment were the only major vendors of VLANs not included in this survey. Although this survey was published in late 1996, not much has changed in these categories.

In the May 1997 test, VLANs from major vendors were tested and Cabletron, Newbridge, and Xylan were judged the "top performers." The reasons given were as follow:

- *Cabletron.* "It works with every type of VLAN and makes management a breeze." It was also praised for routing between VLANs as fast as switching within.

- *Newbridge.* For basing its VIVID switch on MPOA and for having management software that was "among the easiest and most intuitive in the test."

- *Xylan.* Given the accolade of speed and least latency, it also supported every type of VLAN.

The complete list of vendors tested was Agile, Bay, Cabletron, Digital Equipment, Madge, Newbridge, 3Com, and Xylan. This time, while Cisco was again absent (it declined to take part), so was FORE. In general, the magazine found that management was easy and performance was high. All the VLANs included support for high-speed networks, mostly Fast Ethernet, but also FDDI and ATM. This depends on the switches and as mentioned above, it is the type of switch needed that must be chosen first, then the vendor. The issue of the day is, of course, what problems were found.

- *Agile.* Software switching dragged down performance, although Agile was at that moment beta-testing hardware switching. Also, third-party routers were used for inter-VLAN traffic, which could affect performance on heavily loaded networks.

- *Bay Networks.* Low performance and port grouping only—Bay is now about to release better functionality.

- *Cabletron.* While the magazine was hard put to find a fault, it did point out that the switches need reconfiguration of end-station default gateways.

- *Madge.* Doesn't handle multiple VLANs per port, but is praised for performance.

In February 1997 *Communications Week* also ran tests, just on 3Com, Cabletron, and Newbridge. Once again, Cabletron shone, with "A" grades for each category of ease of use, management features, VLAN construction flexibility, and hardware recoverability. The other vendors also scored well, though, on each of these fields, with none getting a "poor" or even only "fair" mark for anything.

John McConnell, a consultant well known for his expertise on VLANs, has done extensive testing in his report *Virtual LANs Head to Head* (see Bibliography). He has evaluated 3Com, Agile, Bay Networks, Cabletron, Cisco, Crosscom, Digital Equipment, FORE Systems, IBM, Newbridge Networks and UB Networks (separately), and Xylan. Highlights of this report include the following points under these categories.

RMON Capabilities.   Only IBM offered none, but all vendors were marked down for not enough support for monitoring. He points out that Newbridge does not support RMON within its VIVID architecture, but the RMON MIB is supported by the VIVID System Manager, Newbridge's network management product. RMON capabilities are left to external probes or embedded agents in hubs at this time. Inconsistent coverage of the basic groups across the product line resulted in significant loss of points for 3Com, coupled with a lack of roving agents. Bay's strength is reflected in the support of the four basic RMON groups across all switch ports, the StackProbe, and conversation steering for some of their switches. Bay's greatest drawback was the lack of an embedded roving agent without the purchase of a management module. UB lost points for not providing the basic monitoring groups for each switch port. Cisco came off worst in this study (lack of roving agents) and FORE best (its ForeRunner products support full RMON per port), as far as monitoring went. Cabletron had a strong score because of its uniform monitoring support with the exception of the ATX switch.

VLAN Definition.   Again IBM scored badly, with support for only MAC-layer bridging within the switch and lack of embedded routing capability. Xylan was given praise for strong network support and for routing methods supported—it received points for supporting distributed route servers and layer 3 switching. 3Com's network layer definitions were noted, but the fact that it does not currently support all the most commonly used protocols within the VLAN architecture was also noted. Agile's strength was deemed to lie in its higher-layer VLANs established by network layer address, protocol, or application. Cabletron's strong network layer support (it supports AppleTalk and DECnet as well as the more common IP/IPX) earned them high scores. Cisco's points were awarded for port- and MAC-based definitions; lack of network layer support lowered its scores. Digital's VLAN strategy was featured for supporting multiple memberships, and its policy-based architecture, which allows a port to be in multiple VLANs, also earned points. Top honors were shared by Agile, Cabletron, Digital, and Xylan, with 3Com not far behind. At the bottom of the heap here were IBM and Bay, with Cisco and Crosscom not far above.

VLAN Management.   Agile's ATMman integration, sharing a common database, launch, icons, and graphical user interface (GUI), was featured. Bay's integration within Optivity and early focus on policies gave it a high score. Cabletron received full scores for spanning tree and multiple backbone support, giving them maximum points; however, all these functions are not available in a single product. Cisco's strengths are reflected in VLAN views and integration with CiscoWorks. Digital's VLAN was found to lack tools for modeling, baselining, and other functions, and this had a strong impact on the score. IBM did not really come into the management picture at all, as it has not clearly defined a management tool yet. Newbridge was praised; integration with the underlying SNMP platform was considered its weakest area. UB also got good marks, with strengths in management views and the tools section. UB's integration was found to be strong with respect to the VLAN applications integrated with the management system. Strengths for FORE were noted in the tools and views sections. Lack of policy-based management features brought its score down, however. Agile's strengths were in the tools and views available through ATMman for Virtual LAN Policy Manager; weaknesses were in policies and, especially, integration levels. Xylan's strengths were found to be in the available toolsets and integration for its VLAN management application; drawbacks again were in its policies (telling, since it emphasizes these) and the VLAN views. 3Com's biggest drawback was also in policy-based management features. All vendors except Xylan and IBM got very high marks in this section.

Conclusion

The McConnell report consistently marks companies down for their poor support of policies. Policies, however, are not yet offered at all by many vendors, and inconsistencies of policy implementation should be viewed while realizing that at least there is something there to criticize.

10

Looking to the Future

Will VLANs Be Taken Up?

As switching becomes universal, bandwidth to the desktop increases, with Fast Ethernet in the end becoming the norm and Gigabit Ethernet adopted in the backbone, VLANs will be seen as an essential part of every switched network. Also, VLANs from different vendors will become more interoperable, encouraging people with heterogeneous networks to climb aboard. However, the need for more dedicated bandwidth will mean that more switches will be needed, making management more complex. Also, the need for flexibility and mobility in the workplace (as mentioned in Chap. 2) means that VLAN groupings will continue to change, again leading to a need for more management.

Again in the name of flexibility, the size of the groups that belong to a particular set of VLANs may become smaller while the number of groups becomes larger, to the point where all users could have a customized mix of services delivered to their desktops. In fact, users could select their own personal mix of services. At present, with VLANs really representing a firefighting strategy to prevent network collapse, this may well be left to the network operating system, but hardware is faster than software and, as policy-based VLANs become more pervasive, the network may take over, especially as the management needs remain the same.

In such a future environment, VLANs change from being a network manager's tool to a business solution, where users subscribe to channels for the applications they need. Given the centralization of applications, where all reside on servers rather than desktops, there will certainly be a need for users to decide what they want, unless technical support staff are to be swamped by constant and daily requests for access to applications. This would also have the benefit of intracompany accounting, with application use within an automated chargeback system for network services. Network managers could still retain control to block access for particular users to specific applications for security purposes.

There is a difference between policy-based network management and policy-based VLANs. Both are important, and—while both are now in the process of development—they are focusing on different values. Policy-based VLANs are one way of making networks more efficient and more secure, but policy-based networking represents a change to a way of business in which VLANs play just a part. Policy-based networking will become increasingly necessary and, as the networking market matures, its use will become critical. Policy-based management is not just VLANs, but management that can be delivered on an end-to-end basis. And no vendor can deliver this over the next 3 years. It will be well into the next millennium before the industry has VLAN functionality for capacity and bandwidth usage policies for customers on the same physical network. Many vendors have articulated the concept but all are far from product delivery. As organizations move toward fully automated, policy-based VLAN structures, network administrators will be able to define access to services with an extremely high degree of precision, establishing specific criteria to be set all the way down to the individual user level or even the time of day. This type of VLAN structure has the added benefit of enabling accurate, automatic tracking of billing and chargeback for network services.

Like policies, standards require compliance. However, policies are higher-level statements than standards, providing general instructions which will last for many years. Standards make specific mention of technologies, methodologies, implementation procedures, and other detailed factors, and thus are likely to last only a few years until conditions change. Even so, the standards will continue to be developed, with a new version of the VLAN standard already under way even though the first iteration is not yet out. The next time around, there must be support for multiple spanning trees, and some of the extra facilities already offered by many vendors must be incorporated.

One further change to the current types of VLANs must become available before takeup is guaranteed—integration with enterprisewide networked directory services such as are already in place in Novell's Intranetware, and soon to be added to Microsoft's NT Server. At present, as mentioned in Chap. 8, the existence of two distinct databases of user lists means that network managers have to struggle to keep them not only up to date but also in sync. With integration, this problem melts away, and will encourage the implementation of both directories and VLANs.

How VLANs Will Change Our Enterprise Networks

Most VLAN discussions focus on the local area, as indeed does this book, but VLANs are also in use in the wide area, as a hidden part of VPNs (virtual private networks) offered by telcos. Several service providers offer LAN services connecting LANs across the WAN for other companies. They need VLANs to separate the traffic and ensure privacy between the different subscribers. The telcos are increasingly using ATM as an intrinsic part of their own infrastructure, and it can be argued that this is where ATM will be most successful, given its origins as a wide area protocol. This would indicate that LANE will be used to provide ELANs for such use, but the better management possibilities have given VLANs the chance to shine, especially where ELANs and VLANs can be linked to provide the greatest possible flexibility. The best VLANs also meet the stringent security needs of the carriers; safeguards are needed against false or duplicate MAC addresses, a safety net which a VLAN adds without complexity.

Within the private (and mainly local) networking world, security will become increasingly tight, and we can expect VLANs to offer user authentication, especially for remote access. As they do so, and as the general level of security rises, VLANs will become an essential part of that security practice and of security policies. For example, management may dictate, through policy, that users must have passwords which are difficult for unauthorized persons to guess. This gives line management the freedom to determine whether to use system-generated passwords or to screen user-selected passwords for difficulty.

New technologies and standards are all very well, but if our businesses are to move forward into the next millennium, they must be based on security and flexibility. The mere containment of broadcast traffic seems very small compared with such large business aims, but virtual LANs have their part to play and, as they move forward, will grow to be a part of the changes in business life. A policy-based approach to networking works to achieve this end. With it, routine and repetitive work can be moved away from network administrators (as is beginning to happen now with some virtual LANs). Also, a policy-based approach is essential to ensure effective secure network management. Security is becoming more and more the focus of enterprise networking, especially as the Internet plays a bigger part, and especially as our workforce becomes more fragmented and flexible. It must be made increasingly clear just who is entitled to what access in the open workplaces of tomorrow.

APPENDIX A

This appendix contains contact details for all the vendors and standards bodies mentioned in the book. Where possible, major regional offices have been included to allow contact no matter what the time zone of the reader.

Agile Networks

Agile Networks, Inc.
1300 Massachusetts Avenue
Boxborough, MA 01719, USA
Tel.: 1-508-263-3600
Fax: 1-508-263-5111

Bay Networks

Corporate Headquarters
Bay Networks, Inc. (formerly SynOptics)
4401 Great America Parkway
Santa Clara, CA 95054, USA
Tel.: 1-408-988-2400
Fax: 1-408-988-5525
http://www.baynetworks.com

Eastern Operations

Bay Networks, Inc. (formerly Wellfleet)
8 Federal Street
Billerica, MA 01821, USA
Tel.: 1-508-436-3680
Fax: 1-508-436-3658

European Regional Offices

Bay Networks
Buropolis, 1240 route des Dolines
06560 Valbonnne Sophia Antipolis, France
Tel.: 33-92-966-966
Fax: 33-92-966-996

Bay Networks
Pompmolenlaan 16
3447 GK Woerden, The Netherlands
Tel.: 31-3480-31611
Fax: 31-3480-31700

Asia-Pacific Regional Office

Bay Networks Hong Kong Limited
Suite 1511-12, 15/F, City Plaza 4/12, Taikoo Wan Road
Hong Kong
Tel.: 852-2-539-1303
Fax: 852-2-523-4028

Cabletron Systems

Corporate Headquarters

Cabletron Systems Inc.
35 Industrial Way, P.O. Box 5005
Rochester, NH 03867-5005, USA
Tel.: 1-603-332-9400
Fax: 1-603-337-2211
http://www.ctron.com

European Regional Office

Cabletron Systems Ltd.
Network House, Newbury Business Park
London Road, Newbury
Berkshire RG13 2PZ, UK
Tel.: 44-1635-580000
Fax: 44-1635-44578

Asia-Pacific Regional Office

85 Science Park Drive, #03-03/04

The Cavendish
Singapore 0511
Tel.: 65-775-5355
Fax: 65-776-3382

Cisco Systems

Corporate Headquarters
Cisco Systems, Inc.
170 West Tasman Drive
San Jose, CA 95134-1706, USA
Tel.: 1-408-526-4000
Fax: 1-408-526-4100
http://www.cisco.com

European, Middle Eastern, and African Headquarters
Cisco Systems Europe, S.A.R.L.
Z.A. de Courtaboeuf
16 Avenue du Quebec, Batiment L2
91961 Les Ulis Cedex, France
Tel.: 33-1-69-18-61-00
Fax: 33-1-69-28-83-26

Japanese Office
Nihon Cisco Systems K.K.
Seito Kaikan 4F
5, Sanbancho Chiyoda-ku
Tokyo 102, Japan
Tel.: 81-3-5211-2800
Fax: 81-3-5211-2810

Digital Equipment

Networking Division Headquarters
Digital Equipment Corporation
550 King Street
Littleton, MA 01460, USA
Tel.: 1-800-457-4825

Fax: 1-508-486-7417
http://www.digital.com

European Regional Office
Digital Equipment Corporation
International (Europe)
12 Avenue des Morgines
Case Postale 176
CH-1213 Petit-Lancy 1, Geneva, Switzerland
Tel.: 41-22-709-4111
Fax: 41-22-709-4140

Asia-Pacific Regional Office
Digital Equipment Asia Pacific Pte Ltd.
300 Beach Road, #39-00
The Concourse
Singapore 0719
Tel.: 65-299-7188
Fax: 65-295-1296

FORE Systems

Corporate Headquarters
FORE Systems, Inc.
175 Thornhill Road
Warrendale, PA 15086, USA
Tel.: 1-412-933-3444
Fax: 1-412-933-6200
http://www.fore.com

EMEA Headquarters
York House, Wolsey Business Park
Tolpits Lane, Watford, WD1 8BL, UK
Tel.: 44-1923-296000
Fax: 44-1923-2966010

Asia-Pacific Headquarters
2115 O'Nel Drive
San Jose, CA 95131, USA
Tel.: 1-408-955-9000
Fax: 1-408-955-9500

Hewlett-Packard Company

Networking Division Headquarters
Workgroup Networks Division
P.O. Box 58059, MS511L-SJ
Santa Clara, CA 95051, USA
Tel.: 1-800-533-1333
Fax: 1-800-533-1917
http://www.hp.com

European Regional Office
Hewlett-Packard SA
150, Route du Nant-d'Avril
CH-1217 Meyrin 2
Geneva, Switzerland
Middle East and Africa tel.: 41-22-780-71-11
European Headquarters tel.: 41-22-780-81-11

Latin America Regional Office
Hewlett-Packard
Latin American Regional Headquarters
Monte Pelvoux No. 111, Lomos de Chapultepec
11000 Mexico, D.F., Mexico
Tel.: 525-202-0155

Japanese Office
Yokagowa-Hewlett-Packard Ltd.
15-7 Nichi Shinjuku 4 Chome
Shinjuku-ku
Tokyo 160, Japan
Tel.: 81-3-3331-6111

Asia-Pacific Regional Office
Hewlett-Packard Asia Pacific Ltd.
17-21/F, Shell Tower
Times Square, 1 Matheson Street
Causeway Bay, Hong Kong
Tel.: 852-599-7777

IBM

Corporate Headquarters
IBM
Old Orchard Road
Armonk, NY 10504, USA
Tel.: 1-914-765-1900
Fax: 1-914-765-4190
http://www.ibm.com

Asia-Pacific Regional Office
3-13 Toranomon, 4 Chome, Minato-ku
Tokyo 105, Japan
Tel.: 81-3-3438-5736
Fax: 81-3-3433-7690

European Regional Office
Tour Pascal, La Défense 7 Sud
Cedex 40 F-92075
Paris La Défense, France
Tel.: 33-1-47-67-60-00
Fax: 33-1-47-67-69-69

Latin America Regional Office
Mt. Pleasant
Route 9
North Tarrytown, NY 10581, USA
Tel.: 1-914-332-2000

Intel

Corporate Headquarters
Intel Corporation
Robert Noyce Building
2200 Mission College Boulevard, P.O. Box 58199
Santa Clara, CA 95052, USA
Tel.: 1-408-765-8080
Fax: 1-408-765-9904
http://www.intel.com

European Regional Offices
Intel Corp. S.A.R.L.
1, Rue Edison, BP 303
78054 Saint-Quentin-en-Yvelines Cedex, France

Intel Corp. (U.K.) Ltd.
Pipers Way
Swindon
Wiltshire, England SN3 1RJ, UK

Intel GmbH
Dornacher Strasse 1
8016 Feldkirchen bei Muenchen, Germany

Asia-Pacific Regional Offices
Intel Japan K.K.
5-6 Tokodai, Tsukuba-shi
Ibaraki, 300-26, Japan

Intel Semiconductor Ltd.
32/F Two Pacific Place
88 Queensway
Central Hong Kong

Madge Networks

Corporate Headquarters
Madge Europe
Knaves Beech Business Park
Loudwater, High Wycombe, Buckinghamshire HP10 9QZ, UK
Tel.: 44-628-858000
Fax: 44-628-858011
http://www.madge.com

North American Offices
Madge Networks, Inc.
2310 North First Street
San Jose, CA 95131-1011, USA
Tel.: 1-408-955-0700
Fax: 1-408-955-0970

Madge Networks
3100 Steeles Avenue East, Suite 803
Markham, Ontario L3R 8T3, Canada
Tel.: 1-905-470-2118
Fax: 1-905-470-9132

Asian Regional Offices
Madge International Ltd.
64-01 Central Plaza
18 Harbour Road
Wanchai, Hong Kong
Tel.: 852-2593-9888
Fax: 852-2519-8022

Madge Japan KK
Believe Mita
43-16 Shiba 3-Chome
Minato-ku, Tokyo 105, Japan
Tel.: 81-3-5232-3281
Fax: 81-3-5232-3208

Israeli Office
LANNET International
Atidim Technological Park, Building No. 3
Tel Aviv 61131, Israel
Tel.: 972-3-645-8458
Fax: 972-3-544-7146

Latin America Regional Office
Avenue Rogue Petroni Junior 999
13th Floor, 04708-000 Morumbi
Sao Paulo, Brazil
Tel.: 55-11-532-2840
Fax: 55-11-532-2888

Newbridge Networks

Corporate Headquarters
Newbridge Networks Corporation
P.O. Box 13600, 600 March Road

Kanata, ON K2K 2E6, Canada
Tel.: 1-613-591-3600
Fax: 1-613-591-3680
http://www.newbridge.com

North and South American Office
Newbridge Networks, Inc.
593 Herndon Parkway
Herndon, VA 22070, USA
Tel.: 1-703-834-3600
Fax: 1-703-471-7080

European, Middle Eastern, and African Offices
Newbridge Networks Ltd.
Coldra Woods, Chepstow Road
Newport NP6 1JB, UK
Tel.: 44-1633-413600
Fax: 44-1633-413615

Asia-Pacific and Former Soviet Union
Newbridge Networks Corp.
P.O. Box 13600, 600 March Road
Kanata, ON K2K 2E6, Canada
Tel.: 1-613-591-3600
Fax: 1-613-591-3680

ODS

Corporate Headquarters
Optical Data Systems, Inc.
1101 East Arapaho Road
Richardson, TX 75081, USA
Tel.: 1-214-234-6400
Fax: 1-214-234-1467
http://www.ods.com

European Offices
Optical Data Systems
Erfurter-Strasse 29, D-85386 Eching
Munich, Germany
Tel.: 49-89-319-5039

Optical Data Systems
Unit 1, Ancells Court
Ancells Business Park, Fleet
Hampshire, GU13 8UY, UK

Plaintree Systems

Corporate Headquarters
Plaintree Systems
Prospect Place, 9 Hillside Avenue
Waltham, MA 02154, USA
Tel.: 1-617-290-5800
Fax: 1-617-290-0963
http://www.plaintree.com

European, Middle Eastern, African Offices
ASMEC Centre
Eagle House, The Ring
Bracknell, RG12 1HB
Tel.: 44-1344-382096
Fax: 44-1344-382095

International Office
59 Iber Road
Stittsville, Ontario, Canada K2S 1E7
Tel.: 1-613-831-8300
Fax: 1-613-831-6120

3Com

Corporate Headquarters
3Com Corporation
5400 Bayfront Plaza, P.O. Box 58145
Santa Clara, CA 95052-8145, USA
Tel.: 1-408-764 5000
Fax: 1-408-764 5001
http://www.3com.com

European Regional Office
3Com Europe Ltd.
Eaton Court, Maylands Avenue
Hemel Hempstead, HP2 7DE, UK
Tel.: 44-1442-278000
Fax: 44-1442-275752

Asia/Pacific Regional Office
3Com Asia Ltd.
Room 2505-07, 25/F Citibank Tower
Citibank Plaza Central
Hong Kong
Tel.: 85-2501-1111
Fax: 85-2537-1149

Whittaker Xyplex

Xyplex Division Headquarters
Xyplex
295 Foster Street
Littleton, MA 01460, USA
Tel.: 1-508-952-4700
Fax: 1-508-952-4704
http://www.xyplex.com

European Office
2 Manor Court, High Street
Harmondsworth, Middlesex, UK
Tel.: 44-181-759-1633
Fax: 44-181-759-1638

South African Office
Xyplex, SA
275 Kent Avenue
Randburg, South Africa
Tel.: 27-11-329-2550
Fax: 27-11-329-2551

Asia-Pacific Regional Office
100 Cecil Street
09-01 The Globe
Singapore 0106
Tel.: 65-225-0068
Fax: 65-225-2050

Xylan

Corporate Headquarters
Xylan Corporation
26679 West Agoura Road
Calabasas, CA 91302, USA
Tel.: 1-818-880-3500
Toll-free: 1-800-999-9526
Fax: 1-818-880-3505
http://www.xylan.com

Asia-Pacific Regional Offices
201 Miller Street, Level 22
North Sydney, NSW 2060, Australia
Tel.: 61-2-959-2267
Fax: 61-2-959-2282

Shinjuku Nomura Bldg. F/32
1-26-2 Nishi-Shinjuku
Shinjuku-ku Tokyo 163-05, Japan
Tel.: 81-3-5322-2956
Fax: 81-3-5322-2929

European Office
Victoria House
Desborough Street
High Wycombe
Buckinghamshire HP 11 2 NE, UK
Tel.: 44-1-494-510993
Fax: 44-1-494-538984

APPENDIX B

STANDARDS BODIES

ANSI (American National Standards Institute)

ANSI
11 West 42nd Street
New York, NY 10036, USA
Tel.: 1-212-642-4900
Fax: 1-212-398-0023
http://www.ansi.org

ATM Forum

World Headquarters
2570 West El Camino Real, Suite 304
Mountain View, CA 94040-1313, USA
Tel.: 1-415-949-6700
Fax: 1-415-949-6705
http://www.atmforum.com

European Office
Boulevard Saint-Michel 78
1040 Brussels, Belgium
Tel.: 32-2-732-8505
Fax: 32-2-732-8485

Asia-Pacific Office
Hamamatsucho Suzuki Building 3F
1-2-11 Hamamatsucho, Minato-ku
Tokyo 105, Japan
Tel.: 81-3-3438-3694
Fax: 81-3-3438-3698

ECMA (European Computer Manufacturers Association)

Rue du Rhone 114
CH-1204 Geneva, Switzerland
Tel.: 41-22-849-6000
Fax: 41-22-849-6001
Telex: 413237 ECMA CH
http://www.ecma.ch

Electronic Industries Association (EIA)

Electronic Industries Association
2500 Wilson Boulevard
Arlington, VA 22201-3834, USA
Tel.: 1-703-907-7500
Fax: 1-703-907-7501
http://www.eia.org

IEEE (The Institute of Electrical and Electronics Engineers)

The Institute of Electrical and Electronics Engineers, Inc.
1838 L Street NW, Suite 1202
Washington DC 20036-5104, USA
Tel.: 1-908-981-0060
Fax: 1-908-981-0027
http://www.ieee.org

IEEE Corporate Office
(New York City)
Tel.: 1-212-705-7900

IEEE Operations Center
445 Hoes Lane
Piscataway, NJ 08855-1331, USA
Tel.: 1-908-981-0060

IEEE European Operations Center (Brussels)
Tel.: 32-2-770-2242
Fax: 32-2-770-8505
Email: memservice-europe@ieee.org

ISO (International Organization for Standardization)

ISO Central Secretariat
1, rue de Varembe
Case Postale 56
CH-1211 Geneva 20, Switzerland
Tel.: 41-22-749-01-11
Fax: 41-22-733-34-30
Email:
   Internet: central@iso.ch
   X.400: c = ch; a = 400net; p = iso; o = isocs; s = central
   http://www.iso.ch
Please note: Copies of ISO standards can be ordered from local standards offices.

ITU (International Telecommunications Union)

ITU
Place des Nations
CH-1211 Geneva 20, Switzerland
Tel.: 41-22-730-51-11
Fax (group 3): 41-22-733-7256
Fax (group 4): 41-22-730-6500
http://www.itu.ch

GLOSSARY*

100Base-F Standard for fiberoptic cabling used with Fast Ethernet, often used to mean Fast Ethernet with fiberoptic cabling.

100Base-T Standard for copper cabling used with Fast Ethernet, often used to mean Fast Ethernet with copper cabling.

100Base-VG LAN standard of the IEE 802 committee for transmission of 100 Mbits/s over UTP (unshielded twisted-pair) cable promoted originally by Hewlett-Packard and AT&T, among others. With 100VG-AnyLAN, an alternative to 100Base-T with Fast Ethernet.

100VG-AnyLAN Method of LAN transmission based on 100Base-VG that builds on aspects of both token ring and Ethernet to run at 100 Mbits/s with resilience and high realization of potential. Generally considered better technically than Fast Ethernet, the alternative, but less successful in the marketplace.

10Base-2 A 10-Mbit/s baseband network using thin Ethernet coaxial cable.

10Base-5 A 10-Mbit/s baseband network using thick Ethernet coaxial cable.

10Base-F The IEEE specification for baseband Ethernet over fiberoptic cabling. (See **10Base-FB, 10Base-FL,** and **10Base-FP.**)

10Base-FB Part of the IEEE 10Base-F specification providing a synchronous signaling backbone that allows additional segments and repeaters to be connected to the network.

10Base-FL Part of the IEEE 10Base-F specification that is designed to replace the Fiber-Optic Inter-Repeater Link (FOIRL) standard providing Ethernet over fiberoptic cabling; interoperability is provided between the old and new standards.

10Base-FP Part of the IEEE 10Base-F specification that allows the organization of a number of end nodes into a star topology without the use of repeaters.

10Base-T IEEE standard enabling telephone UTP cable to be used for Ethernet LANs.

802.1d IEEE standard for spanning tree.

---

*Some definitions in this Glossary are reproduced with permission from Datapro Information Services Group (see Datapro Glossary listed in Bibliography).

802.10   Standard from the IEEE for packet tagging for security within LANs; also used by some companies to tag packets for virtual LANs.

802.11   Unfinished IEEE standard for wireless LANs which use Ethernet bridges with roaming to join them to the network.

802.12   IEEE standard that specifies the physical layer and the MAC sublayer of the data-link layer of the seven-layer OSI reference model.

802.1p   The IEEE standard which adds important filtering controls to 802.1d with VLANs in mind.

802.1Q   The IEEE encapsulation standard, which calls for adding 4 bytes to a packet to tag it for virtual LAN purposes.

IEEE 802.2   IEEE standard for the control of the lower part of layer 2 of the seven-layer OSI reference model.

802.3   IEEE broadband bus networking system that uses CSMA/CD protocol. Ethernet has become the generic name, although it is one trademarked version of 802.3.

802.4   IEEE standard that governs broadband bus and broadband token bus. Usually used in industrial applications.

802.5   IEEE standard that governs token-ring networking systems.

802.6   IEEE standard that governs metropolitan area networks (MANs).

802.9   Integrated voice and data LAN IEEE standard.

AAL5   ATM adaptation layer type 5, part of ATM.

adaptive   A type of bridge or switch which can use adaptive or store-and-forward techniques as needed.

address   Station or user identifier. (See **network address** and **IP address**)

Address Resolution Protocol (ARP)   Internet protocol used to map an IP address to a MAC address.

AppleTalk   Apple protocol consisting of a seven-layer stack similar to the OSI stack.

AppleTalk Address Resolution Protocol (AARP)   Protocol that maps a MAC address to a network address.

ASIC   Application-specific integrated circuit.

ATM   A cell-based networking technique offering high-speed switching together with quality of service.

ATM Forum The main standards-developing body for ATM.

ATM LANE LAN Emulation for ATM; allows logically separated LANs within an ATM network.

ATM route server In a native ATM LAN, the route server addresses the layer 3 requirements, allowing network layer traffic to be routed from one layer 3 domain to another when two ATM LANs are interconnected.

available bit rate (ABR) Quality-of-service class (one of five) defined by the ATM Forum for ATM networks where a timing relationship is not required between sending and receiving stations.

B-ISDN Broadband ISDN; the foundation for ATM.

beatdown The condition a switch suffers when it becomes congested and cells or cell streams representing layer 3 VLAN packets are missed or dropped, leading to the VLAN becoming unstable. Also known as "VC starvation" on an ATM network and "transit VLAN starvation" on a non-ATM network.

bits per second Basic unit of measurement for serial data transmission capacity; abbreviated as kbit/s, for thousands of bits per second; Mbits/s for millions of bits per second; Gbits/s for billions of bits per second; Tbits/s, for trillions of bits per second. The numbers are not exact—kbits/s is really 1024 bits per second, and so on. This is because capacity expands exponentially, not sequentially.

Boostrap Protocol (BOOTP) Allows a computer on the network to act as an address server, automatically giving IP addresses on request.

bridge A relatively simple device that passes data without examining it. Bridges interconnect networks, or network segments, running the same protocols. Operating at the MAC layer, they are protocol-independent—the decision as to whether to forward a signal depends only on the address. There are various types of bridges: encapsulating, translating, and source routing are the main categories.

bridge protocol data units (BPDUs) Formatted frames in spanning-tree networks.

bridge/router Strictly provides bridging at level 2 and routing at level 3, but this very precise meaning has been largely superseded by a looser categorization that includes any device that combines the functions of bridge and router. (See **bridge, brouter,** and **router.**)

broadcast (1) Delivery of a transmission to two or more stations at the same time, such as over a bus-type local network or by satellite. (2) Protocol mechanism in which group and universal addressing is supported.

brouter Strictly speaking, this is a bridge with routing abilities at level 2 (necessary for IBM networks) and source routing at level 2; this very precise meaning has been largely superseded by a looser categorization that includes any device that combines the functions of bridge and router. (See **bridge, bridge/router,** and **router.**)

call admission control (CAC) In ATM, a function that checks to see whether network resources are available to support the QoS and traffic parameters of an incoming call.

cascading hubs Hierarchy of hubs allowing many LAN segments to be connected to a backbone efficiently but without great expense.

category 1 cabling One of five grades of UTP cabling defined in the EIA/TIA-586 standard, used for voice communications; seldom used for data.

category 2 cabling One of five grades of UTP cabling defined in the EIA/TIA-586 standard, used for data transmission at speeds of up to 4 Mbits/s.

category 3 cabling One of five grades of UTP cabling defined in the EIA/TIA-586 standard, used in 10Base-T networks.

category 4 cabling One of five grades of UTP cabling defined in the EIA/TIA-586 standard, used in token-ring networks and transmitting up to 16 Mbits/s.

category 5 cabling One of five grades of UTP cabling defined in the EIA/TIA-586 standard, now becoming the standard for local area networking, and transmitting at up to 100 Mbits/s.

cell In data transmission, a fixed number of bytes of data sent together, as opposed to a frame, which is of variable length.

cell delay variation (CDV) One of three negotiated QoS parameters defined by the ATM Forum.

cell delay variation tolerance (CDVT) Parameter defined by the ATM Forum for ATM traffic management.

cell loss ratio (CLR) One of three negotiated QoS parameters defined by the ATM Forum.

cell relay Transmission technique used in circuit-switching services with fixed-length cells. The main example of cell relay is ATM.

cell transfer delay (CTD) One of three negotiated QoS parameters defined by the ATM Forum.

Challenge Handshake Authentication Protocol (CHAP) Security protocol with encryption that allows access between data communications systems prior to and during data transmission. CHAP uses challenges to verify that a user has access to a system. (See also **Password Authentication Protocol.**)

circuit Means of two-way communication between two or more points.

classical IP over ATM Specification for running IP over ATM as defined in RFC1577. An end station registers its own address with an address server in the ATM network. It then uses the server to learn the ATM addresses of other similar stations. This differs from traditional IP, whose ARP (Address Resolution Protocol) uses broadcasts to learn remote addresses.

client System requesting service from another system.

client/server architecture The division of an application into separate processes capable of operating on separate CPUs connected over a network.

client/sever model In most cases, the "client" is a desktop computing device or program "served" by another networked computing device. Computers are integrated over the network by an application which provides a single system image. The server can be a minicomputer, workstation, or microcomputer with attached storage devices. A client can be served by multiple servers.

connection The establishment of a communication path for the transfer of information according to agreed-on conventions.

connection-oriented service A service where a connection (real or virtual) is set up and maintained for the duration of the call.

connectionless service In a connectionless service, no fixed path is set up between sender and recipient. Every unit of data that is exchanged is self-contained in that it contains all the necessary control and address information to ensure correct delivery, such as packet switching.

constant bit rate (CBR) One of five service categories defined by the ATM Forum for ATM. Where the number of bits sent or received per time unit is constant, or very nearly so.

CSMA/CD   Carrier sense multiple access with collision detection; the method by which Ethernet keeps traffic separated on a network.

cut-through   A type of switching in which the frame is forwarded after the initial header is processed. (See also **store-and-forward switching** and **adaptive.**)

daisychaining   The connection of multiple devices in a serial fashion. An advantage of daisychaining is savings in transmission facilities. A disadvantage is that if a device malfunctions, all the devices daisy-chained behind it are disabled.

data link   Any serial data communications transmission path, generally between two adjacent nodes or devices and without any intermediate switching nodes.

data-link control   (1) Procedures to ensure that both the sending and receiving devices agree on synchronization, error detection and recovery methods, and initialization and operation methods for point-to-point or multipoint configurations. (2) The second layer in the ISO reference model for Open Systems Interconnection.

data-link control layer   Layer 2 of the SNA architectural model, responsible for the transmission of data over a particular physical link. Corresponds approximately to the data-link layer of the OSI model.

data-link layer   The logical entity in the OSI model concerned with transmission of data between adjacent network nodes; it is the second layer processing in the OSI model, between the physical and the network layers.

de facto   According to actual practice; protocols and architectures such as Sun Microsystems' Network File System (NFS) and IBM's Systems Network Architecture (SNA) are called de facto standards because they are so widely used, although they have not been sanctioned by any official standards bodies.

de jure   Standards specified by an accredited standards organization. IEEE 802.3 is an example of a de jure standard.

discrete-event simulation   Modeling technique in which the traffic is represented as sequences of messages, packets, or frames.

domain   A group of nodes on a network forming an administrative entity.

Domain Naming System (DNS)   Used in the Internet for translation of names of network nodes into IP addresses.

Dynamic Host Configuration Protocol (DHCP)   Dynamically allocates IP addresses to end stations for fixed periods of time.

Emulated LAN (ELAN)   Defined by the LANE specification for LANs within an ATM network, there can be multiple ELANs comprising many Ethernets or token rings.

end-system designator (ESD)   Part of GSE for IPv6.

error rate   Ratio of the number of bits, elements, characters, or blocks incorrectly received to the total number of bits, elements, characters, or blocks transmitted.

Ethernet   A local area data network type, developed by Xerox Corporation and supported by all vendors, which uses CSMA/CD. The standard version is IEEE 802.3.

European Telecommunications Standards Institute (ETSI)   An organization made up of national representatives from CEPT countries, the composition of which can include public and private telecommunications providers and equipment manufacturers and users, subject to national determination.

explicit forward congestion indicator (EFCI)   In an ATM network, one of two feedback mechanisms used by the available-bit-rate (ABR) service category to inform the source of the network resources available to it.

explicit rate (ER)   In an ATM network, one of two feedback mechanisms used by the ABR service category to inform the source of the network resources available to it.

explicit tagging   A frame is classified as belonging to a particular VLAN on the basis of a VLAN tag value that is included in the frame.

facilities management   Entering into an agreement with a service supplier to manage internal company facilities such as telecommunications services. Facilities management does not involve the transfer of ownership of facilities to the service provider.

Fast Ethernet   IEEE standard based on IEEE 802.3 but running at 100 instead of 10 Mbits/s and using 100Base-T or 100Base-F families of cabling standards.

fiber distributed data interface (FDDI)   A LAN standard specifying a LAN-to-LAN backbone for transmitting data at 100 Mbits/s over fiberoptic (or copper with TP-PMD) media. Features wrapping rings and includes SMT (surface mounting technology) management.

fiberchannel   Physical layer interface optimized for streaming large volumes of data, developed by ANSI as part of the X3T9.3; now being used for all current Gigabit Ethernet switches.

firewall   Router or access server, often with specialist software, designated as a buffer to ward off intrusion into a private network, particularly important in Internet-to-intranet communications.

first in, first out (FIFO)   A method of transmitting data through a switch to minimize delay. Also has a more general use.

frame   In data transmission, the sequence of contiguous bits bracketed by and including beginning and ending flag sequences.

Generic Attribute Registration Protocol (GARP)   Part of the IEEE 802.1p extension to 802.1d.

GID   GARP Information Declaration.

gigabit   Usually taken to mean 1 billion bits (abbreviated Gbit); more precisely, 1,024,000 bits.

Gigabit Ethernet   Standard approaching IEEE approval for transmission of standard Ethernet traffic at speeds of 1 Gbit/s.

GIP   GARP Information Propagation.

global, site, and end system (GSE)   Alternate addressing architecture for IPv6.

GMRP   GARP Multicast Registration Protocol. Provides a mechanism that allows GMRP participants to dynamically register and deregister information with the MAC bridges attached to the same LAN segment.

graphical user interface (GUI)   A graphics-based front end. Common examples are Microsoft Windows and HP OpenView.

groupware   Networked applications capable of being shared by users.

header   The initial portion of a message, which contains any information and control codes that are not part of the text (e.g., routine, priority, message type, destination addressee, and time of origination).

heterogeneous (computer) network   A system of different host computers, such as those of different manufacturers.

implicit tagging   A frame is classified as belonging to a particular VLAN on the basis of the data content of the frame and/or the receiving port.

International Organization for Standardization (ISO)   Standards body, based in Switzerland, responsible for, among other things, the seven-layer OSI reference model.

Internet A worldwide network of networks all connected using the TCP/IP suite of protocols and functioning as a single virtual network. It provides universal connectivity and three levels of network services: connectionless packet delivery, full-duplex stream delivery, and application-level services (mainly electronic mail).

Internet Activities Board (IAB) Technical policy- and standards-setting body for the Internet, TCP/IP, and connected protocols with two task forces: Internet Engineering Task Force (IETF) and Internet Research Task Force (IRTF).

Internet Control Message Protocol (ICMP) Part of the Internet Protocol, handling error and control messages. (See **Internet Protocol**.)

Internet Control Message Protocol (ICMP) Replaces ARP in IPv6.

Internet Engineering Steering Group (IESG) Organization appointed by the IAB to manage the operation of the IETF.

Internet Engineering Task Force (IETF) Committee which is a subgroup of the Internet Society and is concerned with short- and medium-term problems with TCP/IP and the Internet. It is divided into six subcommittees (with further divisions into working parties). The chairperson sits on the IAB.

Internet Packet Exchange (IPX) A very widely used routing protocol, based on Xerox's XNS, developed by Novell. Implemented in Novell's NetWare.

Internet Protocol (IP) The network layer protocol of the TCP/IP suite including the ICMP control and error message protocol as an integral part. Has 32-bit addressing. Now at version 4. (See **IP next generation** and **Transmission Control Protocol/Internet Protocol**.)

Internet service provider (ISP) A company providing public access to the Internet, usually by leased line, ISDN, and/or modem.

internet(work) Two or more networks linked together. A local Internet is confined within a single building; a campus Internet includes two or more nearby buildings.

intranet A network using Web technologies to communicate within its organization.

IP address Internetwork Protocol address or Internet address. A unique number assigned by an Internet authority that identifies a computer on the Internet. The address consists of four groups of numbers separated by three periods (dots), each between 0 and 255. For example, 195.112.56.75 is an IP address. (See also **IP subnet**.)

IP multicast　Routing technique that allows propagation of IP traffic from one source to many destinations.

IP next generation (IPng)　Colloquial name for IPv6.

IP Security Option (IPSO)　U.S. government specification defining an optional field in the IP packet header for hierarchical packet security levels on a per interface basis.

IP subnet　A division of a network made by allocating different parts of the IP address spectrum to separate the network into IP domains. An IP subnet or subnet mask is a way to subdivide a network into smaller networks to allow a large number of computers on a network with a single *IP address*.

IPv4　The current version of IP now informally redesignated to distinguish it from IPv6.

IPv6　New standard intended to replace IP now being decided by the IETF under RFC1752; offers 128-bit addressing to overcome limitations of numbers of IP addresses. Also has facilities for mobile logins and includes authentication. Also known as *IP next generation* (IPng).

IPX　Internet Packet Exchange.

ISDN　Integrated Services Digital Network; a fully digital communications facility designed to provide transparent, end-to-end transmission of voice, data, video, and still images across the PSTN. Standards for this service are set by the ITU-T.

ITU-T　Part of the International Telecommunications Union.

jitter　Slight movement of a transmission signal in time or phase that can introduce errors and loss of synchronization for high-speed synchronous communications.

Kerberos　Security method for authenticating network users and protecting network traffic based on DES encryption. An Internet Engineering Task Force (IETF) standard, Kerberos works by having a central server grant a "ticket" honored by all networked nodes running Kerberos.

kilobit　Loosely, 1000 bits (actually 1024). Abbreviated kbit.

kilobyte　Loosely, 1000 bytes (actually 1024). Abbreviated kbyte.

LAN Emulation (LANE)　Makes ATM resemble Ethernet or token ring (but not FDDI) to the local network. It allows broadcast and multicast messages. Sitting at the MAC layer, it is protocol-independent—not recognizing the network layer at all.

LAN Emulation Network Node Interface (LNNI) The interface between two LANE servers. LNNI is part of the LANE 2.0 specification, which is expected (at time of writing) to be completed by mid-1997.

LEC LAN Emulation client.

LES LAN Emulation server.

Lightweight Directory Access Protocol (LDAP) An emerging directory access standard for the Internet, also being adopted by NOS vendors.

local area network (LAN) One of the several types of geographically limited communications networks intended primarily for such high-speed data transmission applications as data transfer, text, facsimile, and video.

logical link control (LLC) A protocol developed by the IEEE 802, common to all its local network standards, for data-link-level transmission control. The upper sublayer of the IEEE layer 2 (OSI) protocol that complements the MAC protocol (IEEE 802.2).

management information base (MIB) The set of specifications associated with a vendor's devices, used by management systems.

mapping In network operations, the logical association of one set of values, such as addresses on one network, with quantities or values of another set, such as devices on a second network (e.g., name-address mapping, internetwork-route mapping).

maximum burst size (MBS) In ATM, the number of cells a source is allowed to send at the peak cell rate.

media access control (MAC) A sublayer of layer 2 of the OSI seven-layer reference model depending on characteristics of the underlying physical layer.

metropolitan area network (MAN) High-speed communications network operating within a city or metropolitan area up to 50 km in diameter.

minimum cell rate (MCR) In ATM, the minimum cell rate a source is allowed to maintain.

Multiprotocol over ATM (MPOA) A development for performing switched IP networking over an ATM network, used in conjunction with LANE. Other protocols may be added in the future.

NetBIOS   Network Basic Input/Output System interface, created by Microsoft and IBM.

network   A series of points (nodes, end stations, etc.) connected by communications standards.

network address   A logical address at layer 3 of the seven-layer OSI model, rather than a physical address at layer 2.

network interface card (NIC)   Board that plugs into an expansion slot on a workstation or server which is to be networked, with a connector for the network cabling.

network layer   In the OSI model, the logical network entity that services the transport layer. It is responsible for ensuring that data passed to it from the transport layer is routed and delivered through the network.

network management center (NMC)   Center used for control of a network. May provide traffic analysis, call-detail recording, configuration control, fault detection and diagnosis, and maintenance.

Network Management Forum (NM Forum)   Consortium of over 100 equipment vendors and carriers that are developing implementation specifications for OSI-based network management.

network management system (NMS)   A central software-based set of programs for control of disparate hardware elements; also the whole—both software and hardware used in bridging, routing, etc., together with the network management center.

network-network interface (NNI)   In ATM environments, the interface between two network devices.

network operating system (NOS)   Software that handles the administration of a network to allow resources and files to be shared. Various facilities can be provided, including file sharing, remote access, and a range of administrative functions to control the network.

network topology   Describes the physical and logical relationship of nodes in a network, the schematic arrangement of the links and nodes, or some hybrid combination thereof.

NIC   Network interface card.

node   In a topological description of a network, a point of junction of the links.

non-real-time variable bit rate (nrtVBR)   One of five service categories defined by the ATM Forum for ATM.

Oakley Key Determination Protocol   Protocol that uses a hybrid Diffie-Hellman technique to establish session keys on Internet hosts and routers.

Open Shortest Path First (OSPF)   A link state routing protocol derived from the Dijkstra algorithm. The routers using this protocol update each other and learn network topology by periodically broadcasting link state data across the network.

Open Systems Interconnection (OSI)   ISO's reference model for a seven-layer network architecture used for the definition of network protocol standards enabling all OSI-compliant computers or devices to communicate with each other.

original equipment manufacturer (OEM)   Maker of equipment that is marketed by another vendor, usually under the name of the reseller. The OEM may manufacture only certain components, or complete computers, which are then often configured with software and/or other hardware by the reseller.

OSI seven-layer reference model   An architecture that enables the interoperable transmission of data through a network.

outsourcing   A strategy in which an organization contracts the provision of its requirements for services such as telecommunications, data networking, or information technology to another, separate organization. In outsourcing, the outside organization normally provides the network, equipment, and management necessary to support these services. This contrasts with facilities management, where only the management and maintenance of existing facilities is normally involved.

overlapping VLANs   A virtual LAN type where one user may belong to more than one VLAN at a time.

packet   A group of binary digits, including data and call control signals, that is switched as a composite whole. The data, call-control signals, and error-control information are arranged in a specified format.

Password Authentication Protocol (PAP)   A security protocol without encryption that secures passwords for user authentication to allow access to a network or host. (See also **Challenge Handshake Authentication Protocol.**)

peak cell rate (PCR)   In ATM, the maximum cell rate a source is allowed to maintain.

physical layer   Within the OSI model, the lowest level of network processing, below the data-link layer, that is concerned with the electrical, mechanical, and handshaking procedures over the interface that connects a device to a transmission medium (i.e., RS-232-C).

port　Point of access into a communications switch, a network, or other electronic device.

presentation layer　In the OSI model, the layer of processing that provides services to the application layer, allowing it to interpret the data exchanged, as well as to structure data messages to be transmitted in a specific display and control format.

priority-tagged frame　A tagged frame whose tag header carries no VLAN identification information.

protocol　A collection of rules, voluntarily agreed on by vendors and users, to ensure that the equipment transmitting and receiving data understand each other. In general, protocols represent three major areas: the method in which data is represented or coded (e.g., ASCII); the method in which the codes are received (e.g., synchronously or asynchronously); and the methods used to establish control, detect failures or errors, and initiate corrective action. Terminals performing the same functions under different protocols cannot be used on the same system without protocol converters or emulators.

Protocol Data Unit (PDU)　An OSI term for the data to be transmitted together with the headers and trailers attached by each layer.

protocol-independent multicast (PIM)　Allows the addition of IP multicast routing in an IP network. In dense mode, packets are forwarded on all outgoing interfaces and it is assumed that receivers are most likely to want to receive the packets. In sparse mode, data distribution is limited; packets are sent only if explicitly requested at the rendezvous point.

protocol stack　A defined protocol, together with options applicable for specific functions, which can be implemented as a product. Also called a *functional standard* or *functional profile*.

protocol-transparent.　A device's capability to perform its function independent of the communications protocol.

PVC　In ATM, a permanent virtual circuit.

PVID　Port virtual LAN Identifier.

quality of service (QoS)　A concept by which an application can indicate its specific requirements to the network before transmitting data. Especially applicable to ATM for indicating whether delay is allowed. Soon to be implemented for IP.

real-time variable bit rate (rtVBR) One of five service categories defined by the ATM Forum for ATM.

Remote Authentication Dial-in User Service (RADIUS) Security protocol used to securely transport passwords between the access device and the authentication server.

Remote MONitoring Management Information Base (RMON MIB) An extension of the SNMP MIB II which provides a standards-based method for tracking, storing, and analyzing remote network management information. Developed by the IETF.

request for comments (RFC) Working documents of the Internet research and development community. These documents cover a range of topics related to computer communications ranging from Internet working-group reports to Internet standards specifications.

Resource Reservation Protocol (RSVP) over ATM Will enhance classic IP for obtaining differentiated quality of service over an ATM network.

Reverse Address Resolution Protocol (RARP) TCP/IP protocol for finding IP addresses based on MAC addresses. (See also **Address Resolution Protocol.**)

Revised IP Security Option (RIPSO) RFC1108, which specifies an optional IP header field that contains a security classification and handling label.

RIP Routing Information Protocol.

RMON Standard for Remote Network MONitoring. RMON2 is the latest version.

RMON2 A remote MONitoring standard that complements RMON (RMON1). RMON2 defines network layer and application layer statistics, providing an enterprise view of network traffic.

router A sophisticated, protocol-specific device that examines data and finds the best route for it between sender and receiver. Selects the cheapest, fastest, or least busy of all available routes. Routers are preferable to bridges for large networks with relatively low-bandwidth connections.

routing Assignment of the communications path by which a message or telephone call will reach its destination.

Routing Goop (RG) Identifies the attachment point to the public Internet in GSE for IPv6.

routing table   Table associated with a network node that states for each message (or packet) destination the preferred outgoing link that the message should use.

server   A processor that provides a specific service to the network; for example, a routing server connects nodes and networks of like architectures, and a gateway server connects nodes and networks of different architectures.

Simple Network Management Protocol (SNMP)   A protocol recommended by the IETF for managing TCP/IP networks, internetworked LANs, and packet-switched networks; most commonly employed using TCP/IP protocols. SNMPv2 combines two updates to SNMP: Secure SNMP and Simple Management Protocol; defines everything from SMI to manager-to-manager MIB.

site topology partition (STP)   Corresponds to a site's subnet portion of an IPv4 address, part of GSE for IPv6.

spanning tree   IEEE 802.1d committee standard for bridging LANs without loops. The algorithm ensures that only one path connects any pair of stations, selecting one bridge as the root bridge from which all paths are considered to radiate.

stackable   Network devices that may be stacked vertically but where the stack is managed as one, as opposed to daisychaining, where each device is viewed separately by the network management system. Types of devices available in this format are hubs, switches, and routers.

standard   A document that recommends a protocol, interface, type of wiring, or some other aspect of a network. It may even recommend something as general as a conceptual framework or model (e.g., a communications architecture). De jure standards are developed by internationally or nationally recognized standards bodies or vendors. De facto standards are widely used vendor-developed protocols or architectures, such as IBM's Systems Network Architecture (SNA).

star network   A network topology in which each station is connected to a central station by a point-to-point link and communicates with all other stations through the central station.

store-and-forward switching   A type of packet switching in which the whole frame is processed before further transmission. (See also **cut-through** and **adaptive**.)

sustainable cell rate (SCR)   In ATM, the average cell rate a source is allowed to maintain.

SVC   Switched virtual circuit.

Systems Network Architecture (SNA)   A seven-layer network architecture developed by IBM. Although there is some commonality between the layers of OSI and SNA, a direct mapping is difficult as some layers do not correspond directly.

tag   In VLAN terms, an addition to the header that gives VLAN membership information.

tag control information (TCI)   Part of the VLAN tag header in the IEEE 802.1Q draft standard.

tag protocol identifier (TPID)   Part of the VLAN tag header in the IEEE 802.1Q draft standard.

tag-aware bridges   Bridges that support tagging using the format defined in 802.1Q

tag-aware regions   Regions of a bridged LAN that support tagging using the format defined in 802.1Q

tagged frame   A MAC frame that contains a tag header.

TCP/IP   Transmission Control Protocol/Internet Protocol.

TDM   Time-division multiplexing. Traffic is slotted together along the same pipe with each stream remaining separate.

telecommuter   A work-at-home computer user who connects to the corporate LAN backbone using remote access technologies.

teleconference   Bidirectional audio on either a one-to-many or a many-to-many basis.

telecottage   A well-equipped, well-connected small office that may be used by people working remotely for the same or different companies.

token bus   A local network access mechanism and topology in which all stations actively attached to the bus listen for a broadcast token or supervisory frame. Stations wishing to transmit must receive the token before doing so; however, the next physical station to transmit is not necessarily the next physical station on the bus. Bus access is controlled by preassigned priority algorithms.

token passing   A local area network access technique in which participating stations circulate a special bit pattern that grants access to the communications pathway to any station that holds the sequence; often used in networks with a ring topology.

token ring   A local network access mechanism and topology in which a supervisory frame or token is passed from station to station in sequential order. Stations wishing to gain access to the network must wait for the token to arrive before transmitting data. In a token ring, the next logical station receiving the token is also the next physical station on the ring. The ring may be a logical, rather than physical, ring in the case of switched token ring.

TopN   In *RMON,* conversation information that is imported from an external data collector.

topology   The logical or physical arrangement of stations on a network in relation to one another.

topology   The physical layout of the network including devices and their connections.

TP-PMD   Twisted-pair—physical-media-dependent; an amalgamation of all the old proprietary standards for running FDDI over copper.

transit VLAN starvation   The condition a switch suffers in a non-ATM network when it becomes congested and cells or cell streams representing layer 3 VLAN packets are missed or dropped, causing the VLAN to become unstable. Also known as "beatdown."

Transmission Control Protocol/Internet Protocol (TCP/IP)   A group of transport and network layer protocols designed for wide area networks.

Trivial File Transfer Protocol (TFTP)   A TCP/IP protocol that supports rudimentary file transfer over User Datagram Protocol (UDP). TFTP lacks security controls.

twisted pair   Two insulated wires twisted together; can be shielded (STP) or unshielded (UTP).

twisted-pair—physical-media-dependent (TP-PMD)   ANSI standard for FDDI over copper, replacing SDDI (shielded DDI—runs on shielded cable) and CDDI (copper DDI—runs on unshielded cable).

unspecified bit rate (UBR)   One of five service categories defined by the ATM Forum for ATM.

untagged frame   A MAC frame that does not contain a tag header.

user interface   The program through which the end users interact with the computer.

user network interface (UNI)   In ATM environments, the point at which the user joins the network.

VC starvation   The condition a switch in an ATM network suffers when it becomes congested and cells or cell streams representing layer

3 VLAN packets are missed or dropped, leading to the VLAN becoming unstable. Also known as "beatdown."

VID Virtual LAN identifier.

virtual bridged local area network A bridged LAN in which one or more bridges are tag-aware.

virtual channel (VC) In ATM, a communications track between two nodes giving the bandwidth needed for a virtual connection across the network.

virtual circuit Proposed ITU-T definition for a data transmission service in which the user presents a data message for delivery with a header of a specified format. The system delivers the message as though a circuit existed to the specified destination. One of many different routes and techniques could be used to deliver the message, but the user does not know which is employed.

virtual LAN (VLAN) A network segmented logically, rather than physically, using one of a variety of means at layer 2 or 3 of the OSI model.

virtual path (VP) In ATM, bandwidth between two points on a network used by one or more virtual channels (VCs).

virtual source/virtual destinations (VS/VD) One of three types of feedback an ATM switch can provide.

VLAN Membership Resolution Protocol (VMRP) Protocol that uses GARP to provide a mechanism for dynamic maintenance of the contents of the port egress lists for each port of a bridge, and for propagating the information they contain to other bridges.

VLAN-tagged frame A tagged frame whose tag header carries VLAN identification information.

wide area network (WAN) A network that covers a larger geographic area than a single work site (LAN) or metropolitan area (MAN).

X.500 ITU-T recommendation covering the implementation of addressing databases for devices attached to a network. The basis for Novell's and Banyan's directory services.

BIBLIOGRAPHY

Agile Networks, *Layer-3 Virtual LANs*, white paper, 1996.

Axner, David H., *The Virtual LAN and Network Strategies of Cabletron and Cisco Systems*, technology paper, DAX Associates, 1996.

Bay Networks, *Personal Networking*, white paper, Sept. 1996.

Bellman, Bob, "VLANS: A Look at Three Early Implementers," *Communications Week*, Sept. 1995.

Bennett, Geoff (main author), *Introduction to ATM*, FORE Systems, Jan. 1997.

Bennett, Geoff, *ATM White Paper*, white paper, FORE Systems, 1997.

Cabletron Systems, *SecureFAST VLAN*, white paper, Nov. 1996.

Cisco Systems, *Virtual LAN Communications*, white paper, 1995.

Cisco Systems, *CiscoFusion VLAN Roadmap for Scalable Switched Internetworks*, white paper, 1996.

Corporate Research Foundation, *Corporate Strategies of the Top 100 UK Companies of the Future*, McGraw-Hill, Maidenhead, 1995.

Datapro Information Services, *Glossary*, Feb. 1997.

Demarest, Marc, "The Firm and the Guild: A Perspective on the Future of Knowledge Work and Information Technology," essay, March 1995 (published on the World Wide Web).

Digital Equipment Corp., *Network Switching: Technology, Strategy and Products*, 1995.

Ford, Piers, *LAN Switching: Overview*, Datapro Information Services, Mar. 1996.

Handy, Charles, *The Future of Work*, Basil Blackwell, Oxford, 1984.

Handy, Charles, *The Age of Unreason*, Harvard Business Review Press, Boston, 1992.

Hill, Douglas, *The Switching Book*, Xylan, 1996.

Hofstede, Geert, *Cultures and Organizations*, McGraw-Hill, London, 1991.

IBM, *Switched Virtual Networking from IBM*, white paper, Sept. 1995.

IEEE/ISO/IEC, *Draft Standard for Virtual Bridged Local Area Networks*, IEEE/ISO/IEC Standards Draft, LAN MAN Standards Committee of the IEEE Computer Society, Dec. 1996.

IEEE/ISO/IEC, *P802.1p Standard for Local and Metropolitan Area Networks— Supplement to Media Access Control (MAC) Bridges: Traffic Class Expediting and Dynamic Multicast Filtering,* IEEE/ISO/IEC Standards Draft, LAN MAN Standards Committee of the IEEE Computer Society, Feb. 1997.

Lewis, Lundy, (Cabletron Systems), "Implementing Policy in Enterprise Networks," white paper, *IEEE Communications Magazine,* Jan. 1995.

Mandeville, Robert, and David Newman, "VLANs: Real Virtues," *Datacommunications International,* McGraw-Hill, May 1997.

Mann, Ken, *Lan Internetworking: Overview,* Datapro Information Services, April 1997.

Mann, Ken, *Local Area Networks: Overview,* Datapro Information Services, 1997.

McConnell Consulting, *Virtual LANs: A Reality Check,* white paper, June 1995.

McConnell, John (consultant), *Virtual LANs Head to Head,* analyst report, 1997 (available on Cabletron's Web page).

Mitchell, William J., *City of Bits,* Massachusetts Institute of Technology, Cambridge, Mass., 1995.

Morency, John (of The Registry), "The Demise of the Stand-Alone Router Hails the Rise of VLANs," *Communications Week,* Feb. 1997.

Negroponte, Nicholas, *Being Digital,* Hodder and Stoughton, London, 1995.

Pancucci, Dom, "Virtually Fooled?" *Network News* (VNU), Dec. 1996.

Passmore, David, and John Freeman, *The Virtual LAN Technology Report,* white paper, Decisys, Inc., 1996.

Rash, Wayne, Jr., "The State of the VLAN," *Communications Week,* Feb. 1997.

Raymond, Mark, and Dan Minoli, *ATM and Cell Relay Concepts,* Datapro Information Services, Sept. 1995.

Roberts, Erica, "Virtual LANs?" *Datacommunications International,* McGraw-Hill, Oct. 1996.

Smith, Adam, *The Wealth of Nations,* Henry Regnery Company, Chicago, 1953 (first published in 1776).

Smith, Marina, et al., *ISO Reference Model for Open Systems Interconnection (OSI),* Datapro Information Services, May 1997.

Stevenson, John G., *Switched Virtual Networking Migration,* white paper, IBM, Sept. 1995.

3Com, *A Background to Virtual LANs,* white paper, Dec. 1995.

3Com, *Virtual LAN and ATM Management,* paper and demonstration CD-ROM, Apr. 1996.

Toffler, Alvin, *Power Shift,* Bantam Books, 1990.

Toffler, Alvin, *The Third Wave,* William Collins Sons, 1980.

UB Networks (now part of Newbridge Networks), *ATM White Paper,* white paper, undated.

Also Consulted

World Wide Web pages of all vendors and standards bodies mentioned in the book.

Internet Newsgroups Subscribed To

comp.dcom.cell-relay

comp.dcom.lans.ethernet

comp.dcom.lans.fddi

comp.dcom.lans.token-ring

comp.dcom.lans.misc

INDEX

ABOUT THE AUTHOR

Marina Smith has been at Datapro for more than eight years as an analyst of LANs, internetworking, communications software, network management, and the Internet. She is currently a senior analyst responsible for Datapro's worldwide LANs and Internetworking service. She also acts as advisor to vendors in this field and sometimes writes articles for the trade press on LAN developments.